高校土木工程专业国际化人才培养英文系列教材

Basic Principles of Concrete Structures
混凝土结构基本原理

He Dongqing
贺东青

中国建筑工业出版社
CHINA ARCHITECTURE & BUILDING PRESS

图书在版编目(CIP)数据

混凝土结构基本原理=Basic Principles of Concrete Structures：英文/贺东青主编. —北京：中国建筑工业出版社，2017.12
高校土木工程专业国际化人才培养英文系列教材
ISBN 978-7-112-21648-2

Ⅰ.①混… Ⅱ.①贺… Ⅲ.①混凝土结构-高等学校-教材-英文 Ⅳ.①TU37

中国版本图书馆CIP数据核字（2017）第314312号

This book not only presents the mechanical properties of reinforced concrete materials and methods for the design of individual members for bending, shearing, compression, tension and torsion, but also provides much detail referring to applications in the concrete structural members. An introduction to prestressed concrete is also included.

This book can serve as a textbook for the undergraduates majoring in civil engineering and related majors in colleges and universities, as well as a useful reference book for technicians of civil engineering. We also hope it can help Chinese undergraduate students with a language environment so that they can keep learning English continuously.

责任编辑：吉万旺　牛　松　赵　莉
责任校对：李欣慰

高校土木工程专业国际化人才培养英文系列教材
Basic Principles of Concrete Structures
混凝土结构基本原理
He Dongqing
贺东青

*

中国建筑工业出版社出版、发行(北京海淀三里河路9号)
各地新华书店、建筑书店经销
北京科地亚盟排版公司制版
大厂回族自治县正兴印务有限公司印刷

*

开本：787×1092毫米　1/16　印张：17¾　字数：426千字
2017年12月第一版　2017年12月第一次印刷
定价：**48.00元**
ISBN 978-7-112-21648-2
（31506）

版权所有　翻印必究
如有印装质量问题，可寄本社退换
（邮政编码100037）

Preface

Basic principles of concrete structures is a basic specialty course for undergraduates majoring in civil engineering. This textbook *Basic Principles of Concrete Structures* has three objectives: 1) to establish a firm understanding of the behavior of structural concrete; 2) to develop proficiency in the methods used in current design practice; 3) to help Chinese undergraduate students with a language environment so that they could keep learning English continuously.

The textbook is based on the current national standard *Code for Design of Concrete Structures* GB 50010. The main contents include mechanical properties of steel and concrete, which are very important to mechanical behavior of reinforced concrete structural members. Then step by step, methods for the design of individual members subjected to basic loading types (bending, shearing, compression, tension and torsion) and environment actions are provided. Finally, the knowledge of prestressed concrete structures is stated.

A feature of the textbook is that more emphasis has been laid on basic theories of concrete structures as well as on applications of basic theories both in designing new structures and in analyzing existing structures. The reader will find lots of examples, questions and problems in each chapter, which provide an entry into the literature for those wishing to increase their knowledge through independent study.

The editor in chief of this textbook is He Dongqing from Henan University. Some colleagues have taken part in the editorial work. He Dongqing prepared the draft for Chapters 1,2,5 and 6. Song Pengwei and Fu Xiaoyu prepared the draft for Chapters 3 and 4. Li Fengli prepared Chapter 8. Wu Min prepared Chapter 7 and Yu Mengmeng prepared Chapter 9. Song Pengwei drew all of the figures.

Due to the limited knowledge of the editor, some mistakes and errors in the book may exist. The suggestions for improvement will be gladly accepted.

He Dongqing
October, 2017

Contents

Chapter 1 Introduction ··· 1
  1.1 General Concepts of Concrete Structures ·· 1
    1.1.1 Definition and Classification of Concrete Structures ······················· 1
    1.1.2 Function and Demand of Reinforcement ·· 1
    1.1.3 Advantages and Disadvantages of Concrete Structures ····················· 2
  1.2 Historical Development of Concrete Structures ·· 3
  1.3 Function and Limit State of Structures ··· 4
    1.3.1 Function of Structures ·· 4
    1.3.2 Limit State of Structures ··· 5
    1.3.3 Load and Material Strength ·· 5
  1.4 Characteristics of Course and Learning Methods ····································· 6
  Questions ··· 7

Chapter 2 Mechanical Properties of Concrete and Steel Reinforcement ············ 8
  2.1 Mechanical Properties of Concrete ··· 8
    2.1.1 Strength of Concrete Under Uniaxial Stress State ·························· 8
    2.1.2 Strength of Concrete Under Multiaxial Stresses ··························· 13
    2.1.3 Deformation of Concrete ··· 15
    2.1.4 Fatigue Performance of Concrete ··· 21
  2.2 Mechanical Properties of Steel Bars ·· 23
    2.2.1 Type of Steel Bars ·· 23
    2.2.2 Domestic Ordinary Steel Bars ·· 24
    2.2.3 Strength and Deformation of Steel Bars ····································· 25
    2.2.4 Constitutive Model of Steel Bars ··· 26
    2.2.5 Fatigue Failure of Steel Bars ·· 27
    2.2.6 Reinforcement Properties for Reinforced Concrete Structures ········ 28
  2.3 Bonding Between Steel Bars and Concrete ·· 29
    2.3.1 Importance of Bonding ··· 29
    2.3.2 Components of Bonding Strength ··· 29
    2.3.3 Bond Stress-Slip Relationship ··· 31
    2.3.4 Anchorage of Steel Bars ·· 31
  Questions ··· 34

Chapter 3 Load-Carrying Capacity on Normal Section for Flexural Members … 35
3.1 General Structure of Beams and Slabs … 35
3.2 The Flexural Property of Normal Section for Flexural Members … 37
3.2.1 Three Typical Stages for Beams with Under-Reinforced Beam … 37
3.2.2 Failure Modes … 42
3.2.3 Balanced Failure and Balanced Reinforcement Ratio … 44
3.3 Calculation Principles for Load-Carrying Capacity of Normal Sections … 45
3.3.1 Basic Assumption … 45
3.3.2 Equivalent Rectangular Stress Block … 45
3.3.3 Critical Reinforcement Ratio … 46
3.3.4 Minimum Reinforcement Ratio … 47
3.4 Load-Carrying Capacity of Normal Section for Flexural Members with Singly Reinforced Rectangular Sections … 48
3.4.1 Basic Formulae and Applicable Conditions … 48
3.4.2 Application of Formulae … 49
3.4.3 Calculation Coefficient and Calculation Method on Bearing Capacity of Normal Section … 50
3.5 Load-Carrying Capacity of Normal Section for Flexural Members with Doubly Reinforced Rectangular Sections … 51
3.5.1 Introduction … 51
3.5.2 Basic Formulae and Applicable Conditions … 52
3.5.3 Application of Formulae … 53
3.6 Load-Carrying Capacity of Normal Section for Flexural Members with T-Sections … 56
3.6.1 Introduction … 56
3.6.2 Design Formulas and Applicable Conditions … 58
3.6.3 Application of Formulae … 60
Questions … 63
Exercises … 63

Chapter 4 Load-Carrying Capacity of Oblique Section for Flexural Members … 65
4.1 Introduction … 65
4.2 Inclined Crack, Shear-Span Ratio and Failure Modes of Inclined Sections under Shear Force … 66
4.2.1 Web-Shear Inclined Crack and Flexure-Shear Inclined Crack … 66
4.2.2 Shear-Span Ratio (λ) … 68
4.2.3 Failure Modes of the Oblique Section … 69
4.3 Mechanism of Shear Resistance for Beams without Web Reinforcement … 70
4.4 Calculation of Load-Carrying Capacity on Oblique Section … 72

  4.4.1 Factors Affecting Shear Capacities of Oblique Section ………………… 72
  4.4.2 Design Formulas ……………………………………………………… 74
  4.4.3 Calculation Method of Shear Load-Carrying Capacity of Oblique Section ………………………………………………………………… 79
  4.4.4 Examples ……………………………………………………………… 80
 4.5 Measures to Ensure the Flexural Capacities of Inclined Sections in Flexural Members ………………………………………………………………………… 86
  4.5.1 Moment Capacity Diagram …………………………………………… 86
  4.5.2 Detailing Requirements to Ensure the Flexural Capacities of Inclined Sections with Bent-Up Bars …………………………………………… 88
  4.5.3 Anchorage of Longitudinal Reinforcement at the Supports ………… 91
  4.5.4 Detail Requirements to Ensure the Flexural Capacities of Inclined Sections When Longitudinal Bars are Cut off ………………………… 92
  4.5.5 Detail Requirements of Stirrups ……………………………………… 94
 4.6 Other Structural Requirements of Longitudinal Reinforcement in the Beam and Plate …………………………………………………………………………… 96
  4.6.1 Longitudinal Reinforced Bars ………………………………………… 96
  4.6.2 Handling Reinforcement and Longitudinal Structural Reinforcement ………………………………………………………………………………… 98
 Questions ………………………………………………………………………………… 98
 Exercises ………………………………………………………………………………… 100

Chapter 5 Sectional Load-Carrying Capacity for Compression Members ………… 103
 5.1 Introduction ………………………………………………………………………… 103
 5.2 Details of Compression Member ………………………………………………… 103
  5.2.1 Section Type and Dimensions ………………………………………… 103
  5.2.2 Materials ……………………………………………………………… 104
  5.2.3 Longitudinal Steel Bars ……………………………………………… 104
  5.2.4 Stirrups ………………………………………………………………… 105
 5.3 Compression Capacity of Normal Section for Axial Compression Members ……… 106
  5.3.1 Compression Capacity of Normal Section for Tied Columns ……… 106
  5.3.2 Compression Capacity of Normal Section for Spiral Columns …… 111
 5.4 Failure Modes of Normal Section for Eccentric Compression Members …… 115
  5.4.1 Failure Modes of Eccentric Compressive Short Columns ………… 115
  5.4.2 Failure Modes of Eccentric Compressive Slender Columns ……… 117
 5.5 P-δ Effect of Eccentric Compression Members ……………………………… 120
  5.5.1 P-δ Effect When the Directions of Rod End Moments are Same … 120
  5.5.2 P-δ Effect When the Directions of the Rod End Moments are Different ………………………………………………………………………… 122

5.6 Formulae for Normal-Sectional Compressive Capacity of the Rectangular Sectional Eccentrically Compressed Members ········· 123
    5.6.1 Limit of Large and Small Eccentrically Compressed Failures ········ 123
    5.6.2 Calculation of Normal-Sectional Compressive Capacity for Rectangular Sectional Eccentrically Compressed Members ········· 123
5.7 Calculation of Normal-Sectional Compressed Capacity for the Rectangular Sectional Eccentrically Compressed Members with Asymmetric Reinforcement ········· 129
    5.7.1 Design of Sections with Asymmetric Reinforcement ········· 129
    5.7.2 Evaluation of Ultimate Capacities of Existing Sections ········· 131
5.8 Calculation of Normal-Sectional Compressive Capacity for the Rectangular Sectional Eccentrically Compressed Members with Symmetric Reinforcement ········· 138
    5.8.1 Design of Sections with Symmetric Reinforcement ········· 139
    5.8.2 Evaluation of Ultimate Capacities of Existing Sections ········· 140
5.9 Calculation of Normal-Sectional Compressed Capacity of the I-shaped Sectional Eccentrically Compressed Members with Symmetric Reinforcement ········· 142
    5.9.1 Calculation of Large Eccentric Compression Sections ········· 143
    5.9.2 Calculation of Small Eccentric Compression Sections ········· 144
5.10 N_u-M_u Interaction Diagram and Its Application ········· 147
    5.10.1 N_u-M_u Interaction Diagram of Rectangular Sectional Large Eccentrically Compressed Members with Symmetric Reinforcement ········· 148
    5.10.2 N_u-M_u Interaction Diagram of Rectangular Sectional Small Eccentrically Compressed Members with Symmetric Reinforcement ········· 149
    5.10.3 Features and Applications of N_u-M_u Interaction Diagram ········· 149
5.11 Shear Capacity Formulae for Eccentrically Compressed Members ········· 150
Questions ········· 152
Exercises ········· 153

Chapter 6 Load-Carrying Capacity for Tension Members ········· 155
6.1 Calculation for Cross Section Tension Capacity of Axial Tension Members ········· 155
6.2 Calculation for Cross Section Tension Capacity of Eccentric Tension Members ········· 156
    6.2.1 Large Eccentric Tension Members ········· 156
    6.2.2 Small Eccentric Tension Members ········· 157
6.3 Calculation for Shear Capacity of Eccentric Tension Members ········· 160
Questions ········· 161
Exercises ········· 161

Chapter 7　Load-Carrying Capacity of Torsional Members ········· 162
  7.1　Introduction ··· 162
  7.2　Experimental Study on Members Subjected to Pure Torsion ············ 163
    7.2.1　Performance Prior to Cracking ·· 163
    7.2.2　Performance after Cracking ··· 163
    7.2.3　Failure Modes of Members ·· 164
  7.3　Torsional Capacities of the Cross-section in Pure Torsion ················ 165
    7.3.1　Cracking Torque ··· 165
    7.3.2　Calculation for Torsional Capacities of Members by Variable Angle Space Truss Subjected to Pure Torsion ····························· 166
    7.3.3　Calculation Method for Torsional Capacities of the Members under Torsion in GB 50010 ··· 170
  7.4　Bearing Capacity of Members Under Combined Torsion, Shear and Flexure ··· 173
    7.4.1　Failure Modes ··· 173
    7.4.2　Method of Reinforcement Calculation According to GB 50010 ······ 175
  7.5　Calculation for Torsional Capacities of Reinforced Concrete Frame Column Members with Rectangular Section under Combined Torsion, Shear, Flexure and Axial Force ··· 179
  7.6　Bearing Capacity of Reinforced Concrete Members under Coordination Twist ··· 180
  7.7　Detailing Requirements of the Members under Torsion ··················· 180
  Questions ··· 187
  Exercises ··· 187

Chapter 8　Deflection, Crack and Durability ··································· 188
  8.1　Deflection of Reinforced Concrete Members ································· 188
    8.1.1　Rigidity of Flexural Members ·· 188
    8.1.2　Short-Term Flexural Stiffness B_s ··· 189
    8.1.3　Flexural Stiffness B ··· 192
    8.1.4　Deformation Checking ··· 193
  8.2　Calculation for Crack Width in Normal Sections ···························· 194
    8.2.1　Mechanism of Crack ·· 194
    8.2.2　Average Crack Spacing ·· 195
    8.2.3　Average Crack Width ·· 197
    8.2.4　Maximum Crack Width and Checking ···································· 200
  8.3　Durability of Concrete Structures ·· 202
    8.3.1　Carbonization ·· 203
    8.3.2　Corrosion of Steel Embedded in Concrete ······························ 203

  8.3.3 Durability Design of Concrete Structures ················ 204
 Questions ·············· 206
 Exercises ·············· 207

Chapter 9 Prestressed Concrete Members ·············· 208

 9.1 Introduction ·············· 208
  9.1.1 Basic Concept of Prestressed Concrete ·············· 208
  9.1.2 Classification of Prestressed Concrete ·············· 209
  9.1.3 Methods of Prestressing ·············· 210
  9.1.4 Materials of Prestressed Concrete ·············· 211
  9.1.5 Tension Control Stress σ_{con} ·············· 213
  9.1.6 Prestress Losses ·············· 214
  9.1.7 Combination of Prestress Losses ·············· 223
  9.1.8 Transfer Length of Pretensioned Tendons ·············· 224
  9.1.9 Calculation for Local Compression Bearing Capacity at the End Anchorage Zone of the Posttensioning Members ·············· 225
 9.2 Axial Tension Members of Prestressed Concrete ·············· 229
  9.2.1 Stress Analysis of Members Subjected to Axial Tension ·············· 229
  9.2.2 Calculation for Axial Tension Components at Service Stage ·············· 236
  9.2.3 Checking for Axial Tensile Members at Construction Stage ·············· 239
 9.3 Flexural Members of Prestressed Concrete ·············· 244
  9.3.1 Basic Concept of Balanced Load Design Method ·············· 244
  9.3.2 Stress Analysis of Flexural Members ·············· 246
  9.3.3 Design of Prestressed Flexural Members ·············· 249
 9.4 Detail Requirements of Prestressed Concrete Members ·············· 257
 Questions ·············· 262
 Exercises ·············· 263

Appendix 1 Indexes of Mechanical Properties of Materials as Specified in *Code for Design of Concrete Structures* GB 50010—2010 ·············· 264

Appendix 2 The Nominal Diameter, Nominal Cross-Sectional Area and Theoretical Weight of Nominal Steels ·············· 268

Appendix 3 Environmental Categories of Concrete Structures ·············· 270

References ·············· 271

Chapter 1
Introduction

1.1 General Concepts of Concrete Structures

1.1.1 Definition and Classification of Concrete Structures

The structures mainly made of concrete are called concrete structures, which include plain concrete structure, reinforced concrete structure, prestressed concrete structure and so on. The concrete structures with no reinforcement or non-stressed steel bars are called plain concrete structures, while those with steel bars, steel mesh or steel cage embedded in concrete to carry loads are called reinforced concrete structures. The concrete structures prestressing in concrete by tendons are called prestressed concrete structures. The concrete structures are widely used in industrial and civil buildings, bridges, tunnels, mines, water harbor, etc.

1.1.2 Function and Demand of Reinforcement

Generally, concrete is strong in compression while weak in tension. However, steel is strong in both tension and compression. When the steel reinforcement bars are embedded at the tensile area in a concrete member, they can resist the tension instead of concrete after the concrete cracks. Then steel reinforcement bars and concrete can work cooperatively to provide the resistance of a member.

Taking beams as an example, a plain concrete beam under two concentrated loads is shown in Figure 1-1(a). When the external loads are increased to make the stress in the beam's bottom zone exceed the tensile strength of concrete, the concrete will crack, and the beam will rupture suddenly, indicating the load-carrying capacity of the plain concrete is very low and the failure is brittle. If an appropriate amount of steel bars is embedded at the bottom of the beam, the steel bars will help concrete to sustain the tension due to its good tension-resistant property after the concrete cracks (Figure 1-1b). Thus, the beam can continue to carry loads rather than rupture. The steel bars increase the load-carrying capacity of the beam and improve the beam's deformability, whereby the beam presents obvious warning before its final failure.

As shown in Figure 1-1(c), steels with higher compressive strength are also arranged in axial compression columns to help the concrete sustain compression. Thus, the cross-sectional dimension of the column can be smaller. In addition, the steels in column can also improve the failure load and can sustain the tension by the occasional factors.

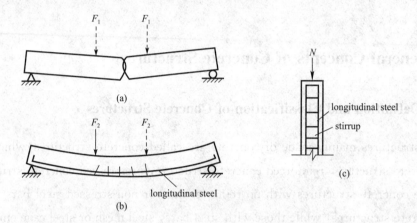

Figure 1-1 Mechanical performance of two beams and an axial compression column
(a) Plain concrete beam; (b) Reinforced concrete beam; (c) Reinforced concrete column

Steel bars can be well bonded to concrete. Thus, they can jointly resist external loads and deform together. The thermal expansion coefficients of concrete and steel are so close ($1.2 \times 10^{-5}/℃$ for steel and $1.0 \times 10^{-5} \sim 1.5 \times 10^{-5}/℃$ for concrete) that the thermal-stress-induced damage to the bond between the two materials can be prevented. At the same time, the arrangement and the number of steel bars should be determined based on correct calculation, detailing requirements and correct construction.

1.1.3 Advantages and Disadvantages of Concrete Structures

Concrete structures have following advantages:

Availability of materials: The materials of concrete used in the largest amount, i.e., sand and gravel are easy to purchase in local market, and industry wastes (e.g., blast furnace slag, and fly ash) can be utilized effectively.

Economy: Reinforced concrete structures take use of the properties of steel and concrete reasonably, and can reduce the cost of construction comparing with steel structure.

Durability: The dense concrete has a high strength. In addition, a thick enough concrete cover can protect the embedded steel from corrosion; thus, the frequent maintenance costs little. So the durability of reinforced concrete is better.

Fire resistance: The steel bar is wrapped in concrete so that it will not quickly reach the softening temperature which leads to structural damage. Compared with wood structure and steel structure, it is better in fire resistance.

Moldability: Concrete structures can be cast according to designed requirements into various shapes and sizes.

Integrity: The members of in situ cast concrete structure are firmly connected to effectively resist dynamic loads such as earthquakes, explosions and other impacts.

Concrete structures also have some disadvantages. Their larger self-weights are disadvantageous for the seismic performance of large span structures and high-rise building structures, and also bring difficulties to transport and construction lifting. In addition, the reinforced concrete structure crack resistance is poor, so most of tensile and bending members in service are behaving with cracks. When a crack is not allowed or there is a strict limit to the crack, prestressed concrete structure should be used. Furthermore, the concrete structures have complicated constructing processes and poor insulation performance.

1.2 Historical Development of Concrete Structures

The concrete structures have a long history of around 150 years. Compared with steel, wood and masonry structures, concrete structures have many advantages in physical mechanical properties, availability of materials, construction costs, etc. So the concrete structures have developed rapidly and have the widest range of applications.

In China, the concrete structure is mostly used in high-rise buildings and multi-storey frames. Concrete-masonry mixing structures are widely used in multi-storey houses. Those structures such as TV towers, water towers, pools, cooling towers, chimneys, tanks and silos are also commonly constructed into concrete and prestressed concrete structures. In addition, concrete structures are also common in large-span public buildings and industrial buildings.

At present, the tallest reinforced concrete building in the world is Khalifa Tower, with a height of 828m, located in Dubai, the United Arab Emirates. The prestressed concrete television tower in Toronto, Canada, with a height of 549m is a representative prestressed concrete structure. The world's tallest concrete gravity dam is Grande Dixence Dam in Switzerland, with a height of 285m and a width of 15m at the top of the dam, 225m at the bottom of the dam and 695m in length. China's Three Gorges Project of Yangtze River is the largest water conservancy project in the world. The concrete dam is 186m high and the amount of concrete is 15,270,000mm^3.

Waterlocks, hydroelectric dams, docks and wharves built with reinforced concrete are also very extensively applied in China.

In recent years, China has made a lot of new achievements in the research of concrete basic theory and design methods, structural reliability and load analysis, industrial building system, structural seismic and finite element analysis methods, and modern test technology. Advanced modern test technology ensures that experimental research is more accurate and systematic. The analysis method based on reliability theory is also gradually improved, and begins to use in the whole structure and during the whole process. At the same time, the period of structural design has been shortened by the popularity and multifunction of electronic computers, the development of CAD and other software systems.

In addition, we have made great progress in the design theory and methods of concrete structure. The current specification *Code for Design of Concrete Structures* GB 50010—2010, which has accumulated rich engineering practices and scientific achievements over the past more than half a century, and improved the design method of concrete structure in China to the present international level. And it plays a guiding role in engineering design.

With the development of high strength steel bars, high performance admixtures and hybrid materials, the high performance concrete has been used widely, and the research and the application of steel fiber concretes and polymer concretes have made great development. Besides, lightweight concrete, aerated concrete, ceramic concrete and green concrete using industrial waste slag not only improve the performance of concrete, but also have important significance for energy conservation and environmental protection. Concrete for special needs such as ray-resisting, hard-wearing, corrosion-resisting, anti-osmosis, heat preservation and intelligent concrete with its structure are also in development.

1.3　Function and Limit State of Structures

1.3.1　Function of Structures

In order to ensure that the structure of the design is safe and reliable, the building structure should meet the requirements of its function. The function of the structures includes three aspects: safety, applicability and durability. Safety refers to the reliability of the bearing capacity of building structures. The structure should be able to withstand all kinds of loads and deformations during the normal construction and service, and the overall stability of the structure can be maintained during and after an earthquake, explosion, etc. The applicability requires that the structure cannot produce too much deformation and cannot have wide cracks and vibrations during the normal service period. The durability requires that the structure will not have serious weathering, corrosion, peeling and carbonization in the normal maintenance condition, so as to achieve the expected life of the design.

1.3.2 Limit State of Structures

The entire structure or part of a structure exceeding a certain state will not meet the design requirements of a function and this state is called the limit state of this function. So the limit state of structures is the state of a boundary between reliability and failure.

The limit state of structures can be divided into the ultimate limit state and the serviceability limit state.

1. Ultimate limit state

The ultimate limit state is when the structure or member reaches the maximum carrying capacity or deformation is not suitable for the continuous bearing state. The structure under such conditions including damages by lack of material strength, the fatigue damage, too much plastic deformation to bear load, loss of stability or turning into mechanism system, is considered to exceed the ultimate limit state. The structure or member cannot meet the requirements of security after exceeding the capacity limit state.

2. Serviceability limit state

The structure or member reaches a specified limit in the service or durability of a structure or member is called serviceability limit state.

For example, when the structure or member appears excessive deformation, excessive cracks, local damage, and vibrations that affect normal use, it can be considered that the structure or member exceeds the serviceability limit state. Beyond the serviceability limit state, the structure or member cannot guarantee the applicability and durability of the functional requirements.

When structural design is carried out, the structure or member should be calculated according to the ultimate limit state, and also be checked according to the serviceability limit state. In other words, the structure or member should satisfy the requirements of the ultimate limit state while satisfying the serviceability limit state.

1.3.3 Load and Material Strength

The value of a load that does not change over time is called a permanent load or a constant load (expressed as G or g), such as the self-weight of the structure. The load that varies with time is called a variable load or a live load (expressed as Q or q), such as floor live load.

The standard value of a load is its basic representative value denoted by the subscript k, which are used to check the deformation and the crack width of the concrete structure. The design value of a load should be adopted in order to meet the reliability requirement of structures when calculating the load-carrying capacity of the section. The design value is equal to the standard value multiplied by the load factor. The load factor of a constant load γ_G is generally 1.2. The load factor of a live load γ_Q is generally 1.4.

The internal force obtained by the load standard value is called the standard value of internal force, such as the standard value of the bending moment M_k and the axial force standard value N_k. The internal force, which is calculated by load design value, is called the design value of internal force, such as the design value of bending moment M and the axial force design value N.

The standard value of material strength should be adopted when checking the structural deformation and crack width. When calculating the load-carrying capacity of the section, the design value of material strength should be used. The design value of material strength is equal to the standard value divided by the material sub-coefficient. For example, the design value of axial tensile strength of concrete (f_t) is equal to the standard value ($f_{t,k}$) divided by the material sub-coefficient (γ_c).

1.4 Characteristics of Course and Learning Methods

This course is one of the fundamental specialized courses for undergraduate students majoring in civil engineering. From the study of this course, students should know the basic mechanical properties, computational analysis methods and detailing of structural members composed of concrete and reinforcement, understand the distinctions and similarities between this course and previous mechanics courses, acquire the ability of solving real engineering problems in structural design and assessment, and lay a solid foundation for future design courses. To study this course more effectively, students should do the following:

(1) Pay attention to the differences and similarities between this course and previous courses, especially that the basic principles of concrete structures are equivalent to the mechanics of reinforced concrete and prestressed concrete.

(2) Concrete structural theories are mostly based on the experimental research. There has not been a complete or generally accepted theoretical system up to now. Therefore, students should pay attention to the site visit and understand the actual projects.

(3) To ensure the safety and reliability of structures, only undertaking quantitative theoretical analysis is not enough. Qualitative detailing measures are necessary as well.

These measures are summaries of previous experiences. Although they cannot be explained quantitatively, profound principles are behind them. So in the study, students should understand the meaning of the detailing measures rather than just memorize them.

(4) Study fundamental theories with the goal of applying them in future engineering practice.

(5) Combining theory with practice is helpful in the study of this course.

Questions

1-1 What is reinforced concrete structure? What are the main roles and requirements of reinforcement?

1-2 What are the main advantages and major drawbacks of reinforced concrete structures?

1-3 What are the functional requirements of the structure? Briefly describe the concepts of the ultimate state and the serviceability limit state.

1-4 What are the main characteristics of this course? What are the issues that need to be addressed in this course?

Chapter 2
Mechanical Properties of Concrete and Steel Reinforcement

2.1 Mechanical Properties of Concrete

2.1.1 Strength of Concrete Under Uniaxial Stress State

Although concrete structures or members usually work under complex stress state, the strength of concrete under uniaxial stress state is the foundation and significant index of the strength of concrete under complex stress state.

The size and shape of specimen, test method and loading rate will affect the test results. Therefore, uniform standard test methods are conducted domestically and abroad.

1. Concrete under uniaxial compression

(1) Compressive strength of concrete cube

Compressive strength of the standard cube is the basic strength index of concrete in Chinese code because of its stabilization. At present in *Code for Design of Concrete Structures* GB 50010, the grades of concrete are classified by the compressive strength of cubic specimens with side length of 150mm under standard curing conditions for 28 days (the temperature is at 20 ± 3°C, the humidity is over 90%). The compressive strength measured according to the standard test method is adopted as the concrete cube compressive strength, in unit of "N/mm^2".

As stipulated in the *Code for Design of Concrete Structures* GB 50010, concrete strength grade is determined by the characteristic value of the cube strength (denoted by $f_{cu,k}$). Subscript "cu" means cube, and "k" means characteristic value. Grades are determined with the assured ratio not less than 95% by the probability analysis based on the test values of standard cubic strength. The cubic strengths of concrete are classified into the following grades: C15, C20, C25, C30, C35, C40, C45, C50, C55, C60, C65, C70, C75, C80. For example, concrete with the strength grade of C30 means its characteristic strength $f_{cu,k}$ is equal to 30N/mm^2. Concretes of C50~C80 are defined as high-strength concrete.

The strength grade of concrete in reinforced concrete structures should not be lower than C20 and should not be lower than C25 for structural members using reinforcement of 400MPa or larger. For prestressed concrete structures, C30 is the minimum and C40 or larger is preferred.

The test method has a great influence on the compressive strength and the failure mode of cubic concrete specimens. After loose concrete is removed, the failed specimen looks like two pyramids connected top against top. This is because the specimen will shorten vertically and expand laterally when subjected to vertical compression, but the friction exists between the specimens and the loading plate under a multi-axial loading state, just as if the specimen were restrained by two hoops at the ends (Figure 2-1a). If the top and bottom surfaces of the specimen are greased, the friction between the specimen and the loading plates is significantly reduced. The specimen is almost under a uniaxial compression state. The restriction against lateral expansion of the specimen is approximately constant along the specimen height. Cracks paralleling to the loading direction can be observed, and the measured strength is lower than that of the ungreased specimen (Figure 2-1b). Specimens should not be greased according to the standard test method in China.

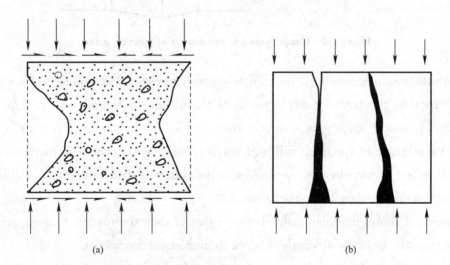

Figure 2-1 Damage of concrete cube test block
(a) Non-lubricant; (b) Lubricant coating

For cubic concrete specimens, the faster the loading rate is, the higher the measured strength will be. The loading rate is generally specified as $0.3 \sim 0.5 \text{N/mm}^2$ per minute for concrete specimens with a cube strength lower than 30N/mm^2 and $0.5 \sim 0.8 \text{N/mm}^2$ per minute for concrete specimens with cube strength equal to or greater than 30N/mm^2. The cube strength of concrete will increase with its age, 28 days is chosen as the test standard.

(2) Compressive strength of concrete prism

The strength of concrete is also related with the shape of specimen. The axial compressive strengths measured on prismatic specimens can obviously reflect the real compression capacity of concrete better than the cube strength.

Standard for test method of mechanical properties on ordinary concrete stipulates that the standard specimen for axial compressive strength should be a prism in the dimension of 150mm×150mm×300mm. The fabrication condition for prismatic specimens is the same as that of cubic specimens, and neither prismatic nor cubic specimens are greased. The compression test of the prism and the failure of the specimen are shown in Figure 2-2.

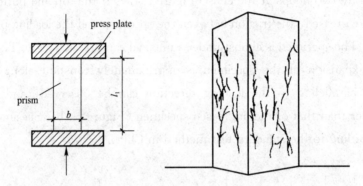

Figure 2-2　Compression test and damage of concrete prism

With the height increasing, the friction between specimen and loading plates will have less influence on the laterally displacement at the middle of height, so the compressive strength of prismatic specimens is smaller than cube's. The bigger the aspect ratio is, the smaller the strength of specimen will be. However, when the aspect ratio reaches a certain value, the effect is not obvious. According to the collected data, the effect is eliminated when the ratio is between 2~3 approximately.

Figure 2-3 shows the relationship between part of the experimental data on axial compressive strength and cube strength obtained in domestic experiments.

It can be assumed that within a certain range, the axial compressive strength f_{ck} is approximately proportional to the cube strength $f_{cu,k}$. Based on experimental research, *Code for Design of Concrete Structures* GB 50010 conservatively expresses the relation between the two strengths as follows:

$$f_{ck} = 0.88\alpha_{c1} \times \alpha_{c2} \times f_{cu,k} \tag{2-1}$$

Where　α_{c1}——the ratio of the prism strength on the cube strength; α_{c1} equals 0.76 for concrete with the strength grade lower than C50 and α_{c1} equals 0.82 for C80 concrete; the value of α_{c1} is linearly interpolated between 0.76 and

0.82 for C55~C75 concrete;

α_{c2} ——the reduction coefficient considering the brittleness of high-strength concrete; α_{c2} equals 1.0 for C40 concrete and α_{c2} equals 0.87 for C80 concrete; the value of α_{c2} is linearly interpolated between 1.0 and 0.87 for intermediate grades;

0.88 ——a parameter to consider the strength differences between laboratory specimens and real structural members due to different fabrication methods, curing conditions and loading states.

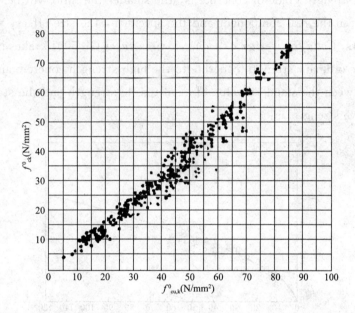

Figure 2-3 The relationship between the axis compressive strength of the concrete and the compressive strength of the cube

In some countries or regions, cylinder specimens are chosen to determine the axial compressive strength of concrete. For example, concrete cylinders with the diameter of 6in (152mm) and the height of 12in (305mm) are adopted as the standard specimens for the axial compressive strength in America, Japan and CEB (European Committee for Concrete), referred to as f'_c. For the concrete below C60, the relationship between the compressive strength of the cylinder f'_c and the standard value of cube compressive strength $f_{cu,k}$ can be calculated according to Eq. (2-2). When $f_{cu,k}$ is more than C60 with the increase of compressive strength, the ratio between f'_c and $f_{cu,k}$ is also increased. CEB-FIP and MC-90 are specified that for C60 concrete, the ratio is 0.833; for C70 concrete, the ratio is 0.857; for C80 concrete, the ratio is 0.875.

$$f'_c = 0.79 f_{cu,k} \tag{2-2}$$

2. Concrete under uniaxial tension

(1) Tensile strength of concrete

Tensile strength is one of the basic mechanical indexes, and its standard value is defined as f_{tk}, where 't' means tensile and 'k' means characteristic value. The axial tensile strength can be obtained from the axial tensile test.

Figure 2-4 shows the results of the concrete axial tensile strength test. It can be seen that the axial tensile strength is only 1/17~1/8 of the compressive strength of the cube. The higher the strength grade of concrete is, the smaller the ratio will be. Integrating the difference between the real component and the specimen, the size effect, the acceleration and other factors, *Code for Design of Concrete Structures* GB 50010 takes into account the change from the ordinary strength concrete to the high strength concrete and proposes the relationship between the standard value of axial tensile strength and the standard value of cube compressive strength as:

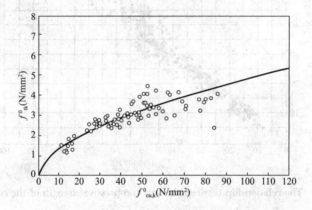

Figure 2-4 The relationship between the axis tensile strength of concrete and the compressive strength of cube

$$f_{tk} = 0.88 \times 0.395 f_{cu,k}^{0.55} (1 - 1.645\delta)^{0.45} \times \alpha_{c2} \tag{2-3}$$

Where δ——coefficient of variation.

The meaning of 0.88 and the value of α_{c2} are the same as Eq. (2-1).

Where the coefficient 0.395 and 0.55 are the reduction factors between the axial tensile strength and the cube compressive strength.

(2) Splitting tensile strength of concrete

It is more difficult to measure the tensile strength than the compressive strength because of testing machine limitations. A number of methods are the cylinder (or prism) splitting test and the direct pulling test as shown in Figure 2-5.

According to the elastic theory, the axial tensile strength test value f_t^0 (superscript "0" indicates the test value) can be calculated according to the following equation:

$$f_\mathrm{t}^0 = \frac{2F}{\pi dl} \qquad (2\text{-}4)$$

Where d——the side length of the cubic prism or the diameter of the cylinder;

l——the length of cubic prism or the length of the circular cylinder;

F——failure load.

Experimental results show that splitting tensile strength is only slightly higher than the tensile strength, and the size of the split specimen also has a certain effect on the experimental results.

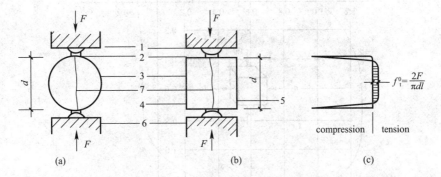

Figure 2-5 Diagram of concrete splitting test

(a) Splitting test with a cylinder; (b) Splitting test with a cube; (c) Horizontal stress distribution in split surface

1—The above press platen; 2—Curved pad and cushion; 3—Specimen; 4—Pouring on the top surface of mold;
5—Pouring on the bottom surface of mold; 6—The bottom press platen; 7—The crack line of specimen

2.1.2 Strength of Concrete Under Multiaxial Stresses

1. Concrete under biaxial stress

In structural members, concrete is usually subjected to multi axial stresses rather than the ideal uniaxial loading. For example, beams always not only carry moment, but also bear shear force and axial force. Therefore, investigating the multi axial behavior of concrete is of great importance for better understanding the properties of concrete members and improving the design and research of concrete structures.

The failure curve of concrete under a biaxial stress state (Figure 2-6) can be obtained by applying normal stresses σ_1 and σ_2 in two mutually perpendicular directions while keeping the normal stress at zero in the third direction perpendicular to the aforementioned two directions and recording the strengths of concrete under different stress ratios (σ_1/σ_2).

It can be seen that under biaxial tension (the first quadrant in Figure 2-6), σ_1 and σ_2 do not greatly influence each other, and the biaxial tension strength of concrete is approximately equal to the uniaxial tensile strength. However, for concrete under biaxial compression, the strength in one direction increases with the buildup of compressive stress in another direction (the third quadrant in Figure 2-6). The biaxial compressive strength can be as much as 27% higher than the uniaxial strength under the combined loading of tension and compression that reduces both the tensile and compressive stresses at failure (the second and the fourth quadrant in Figure 2-6).

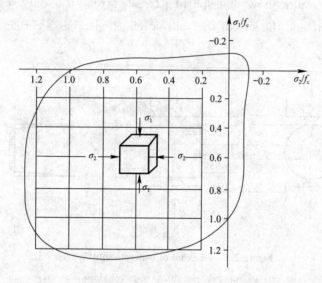

Figure 2-6 Damage envelope diagram of concrete in bidirectional stress state

Figure 2-7 shows the failure curve of concrete under combined normal and shear stresses. It is found that the shear strength of concrete will increase with the buildup of compressive stress when the latter is small. But after the compressive stress exceeds about $0.6f_c$, the shear strength will decrease with the increase of compressive stress. On the other hand, the existence of shear stress reduces the compressive strength of concrete. Similarly, the shear strength decreases with the increase of tensile stress. And the tensile strength of concrete reduces in the presence of shear stress.

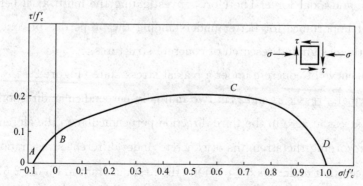

Figure 2-7 The failure curve of the combination of stress and shear stress

A—axial tension; B—pure shear; C—shear-compression; D—axial compression

2. Concrete under triaxial stress

The curve of axial stress-strain under triaxial compression is shown in Figure 2-8. The test is conducted as follows: keep the confining fluid pressure during loading process and increase axial stress until specimen is broken, then measure the displacement of axial direction.

It can be seen that the strength and strain are both improving with the increase of confining stress. Researches have shown that for axially loaded concrete cylinders, the axial strength will be greatly improved subjected to uniform confining fluid pressure. The increased amplitude is approximately proportional to the confining pressure (Figure 2-8).

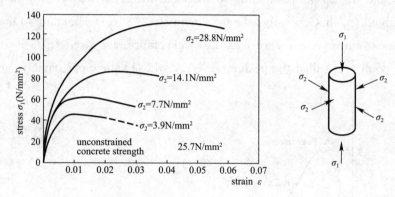

Figure 2-8 The axial stress-strain curve of the concrete cylinder applied to the compression test

When σ_2 is not very large, the ultimate compressive strength f_{cc} in the direction of σ_1 can be expressed as:

$$f'_{cc} = f'_c + (4.5 \sim 7.0) f_L \tag{2-5}$$

Where f'_{cc} ——the axial compressive strength of specimen with side pressure;

f'_c ——the compressive strength of the unconfined concrete cylinder;

f_L ——the confining pressure; The number before f_L is the coefficient of confining pressure, which the average equals 5.6, a bit higher value is suggested when the confining pressure is low.

Based on the aforementioned mechanism, the load-carrying capacity and deformation property of concrete columns subjected to compression can be improved by the dense arrangement of circular hoops or spirals at the column periphery to retrain the lateral deformation of internal concrete. The performance of concrete filled steel tube is similar with the former.

2.1.3 Deformation of Concrete

1. Stress-strain relation of concrete under a short-term load

(1) Stress-strain curve of concrete in uniaxial compression

Figure 2-9 illustrates a typical measured full stress-strain curve of concrete under axial compression, which can be divided into two parts, i.e., the ascending branch (OC) and the descending branch (CF). In the ascending branch, segment OA is approximately a straight line, which is mainly due to the elastic deformation of aggregates and cement crystals while the influence of the viscous flow of hydrated cement paste and the evolution of initial micro-cracks are small. With the increase of stress, the ascending slope of the curve gradually decreases due to the viscous flow of unhardened gel in concrete and the propagation and growth of micro-cracks. When the stress is increased nearly to the axial compressive strength, large strain energy is stored in the specimen, internal cracks speed up their propagation, and the cracks paralleling to the axial load link together, which means the specimen is about to fail. Generally, the maximum stress σ_0 corresponding to the peak point C in the stress-strain curve is regarded as the axial compressive strength f_c of concrete, and the strain at point C is called the peak strain(ε_0), whose value approximates at 0.002.

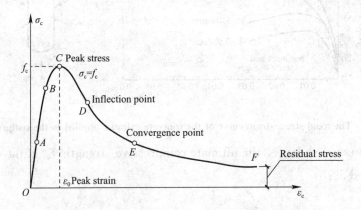

Figure 2-9　Compressive stress-strain curve of concrete prism

After point C, further development and connection of continuous cracks damage the prismatic specimen more and more severely, causing the specimen to lose its capacity gradually as shown by the descending branch CF in Figure 2-9. The descending branch is hard to record by ordinary test machines because with the decrease of stress, the strain energy stored in the machine is released and the sudden recovery deformation of the machine will crush the specimen with severe damages. Therefore, machines of large stiffness or with certain auxiliary devices must be adopted, and the strain rate should be strictly controlled so as to obtain the descending branch of the stress-strain curve.

(2) Mathematical models of stress-strain relationship of concrete under uniaxial compression

There are two mathematical models to describe the stress-strain relationship of concrete under axial loading.

1) Model suggested by Hognestad

The model suggested by Hognestad assumes the ascending branch and the descending branch as a second-order parabola and an oblique straight line (Figure 2-10), respectively, expressed by Eqs. (2-6) and (2-7).

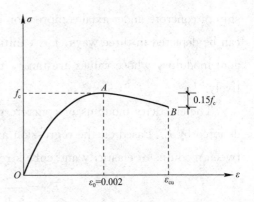

$$\sigma = f_c \left[2 \frac{\varepsilon}{\varepsilon_0} - \left(\frac{\varepsilon}{\varepsilon_0} \right)^2 \right] \quad (2\text{-}6)$$

$$\sigma = f_c \left[1 - 0.15 \left(\frac{\varepsilon - \varepsilon_0}{\varepsilon_{cu} - \varepsilon_0} \right)^2 \right] \quad (2\text{-}7)$$

Figure 2-10 The stress-strain curve proposed by Hognestad

Where f_c ——the peak stress (the axial compressive strength of concrete);

ε_0 ——the strain corresponding to the peak stress, taking the value as 0.002;

ε_{cu} ——the ultimate compressive strain, taking the value as 0.0038.

2) Model suggested by Rüsch

The model suggested by Rüsch also adopts a parabolic ascending branch, but the descending branch is a horizontal straight line (Figure 2-11), expressed as:

$$\sigma = f_c \left[2 \frac{\varepsilon}{\varepsilon_0} - \left(\frac{\varepsilon}{\varepsilon_0} \right)^2 \right] \quad (2\text{-}8)$$

$$\varepsilon = f_c \quad (2\text{-}9)$$

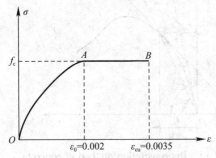

Figure 2-11 The stress-strain curve proposed by Rüsch

(3) Stress-strain curve of concrete in axial tension

Due to the difficulty to obtain the stress-strain curve under tension, there is little experimental data up to now. Figure 2-12 is the stress-strain curve of the concrete under axial tension measured by electro-hydraulic servo testing machine. The shape of the curve is similar to that under compression, with ascending and descending segments. Experiments show that at the initial stage of specimen loading, the deformation and stress increase linearly to 40%~50% of the peak stress, reaching the limit of proportionality. When it is loaded to 76%~83% of the peak stress, the curve has a critical point (i.e., the starting point of the unstable fracture propagation). The corresponding strain is only $75 \times 10^{-6} \sim 115 \times 10^{-6}$ when it reaches the peak stress. The slope of the curve at the descending section is steeper with the increase of concrete strength grade. The value of tensile elastic modulus and the value of compressive elastic modulus are basically the same.

(4) Elasticity modulus of concrete

Modulus of concrete is the ratio of stress to strain. Because the stress-strain relationship of concrete under axial compression is a curve, the modulus of concrete is a variable. It can be depicted in three ways, i. e., initial modulus of elasticity, secant modulus and tangent modulus, whose values are $\tan\alpha_0$, $\tan\alpha_1$, and $\tan\alpha$ as shown in Figure 2-13, respectively.

The elasticity modulus of concrete generally means the initial modulus of elasticity, denoted by E_c. Based on the regression analysis of experimental data, the relationship between modulus of elasticity and cube strength can be obtained as follows:

$$E_c = \frac{10^2}{2.2 + \frac{34.7}{f_{cu,k}}} (\text{kN/mm}^2) \tag{2-10}$$

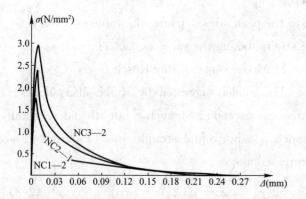

Figure 2-12　Tensile stress-strain curve of concrete with different strengths

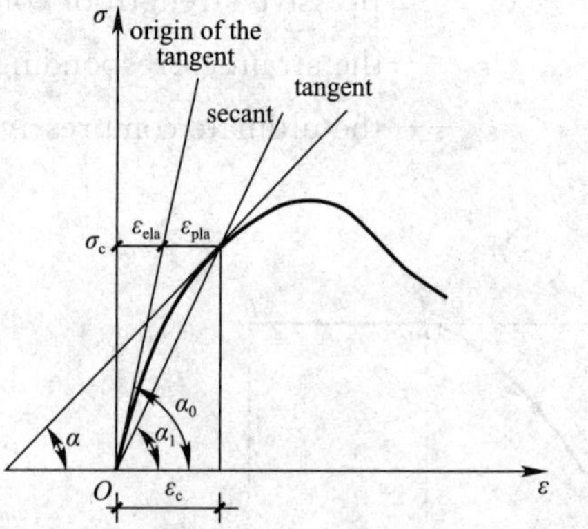

Figure 2-13　Representation method of concrete deformation modulus

The elasticity modulus given by *Code for Design of Concrete Structures* GB 50010 is listed in Appendix 1. It is of practical meaning to use secant modulus or tangent modulus in nonlinear analysis of concrete structures, because the two modulus can better reflect the characteristics of stress-strain relationship curves. And the relation between the secant modulus E_c' and the initial modulus of elasticity E_c can be expressed as:

$$E_c'' = \upsilon E_c \tag{2-11}$$

Where　υ——the coefficient of elasticity, decreasing with the increase of stress.

2. Deformation properties of concrete under long-term load

Under long-term loading, creep is the increment in strain with time due to a sustained load.

Figure 2-14 illustrates the creep curve of a prismatic concrete specimen. An instantaneous strain ε_c will be recorded when the specimen is loaded to a certain value of stress (as $0.5 f_c$ in Figure 2-14). If the stress is kept constant, the specimen deformation will continuously increase with time, expressed as the creep strain ε_{cr}. In the first several months of loading, the creep strain increases rapidly and 70%~80% of the total creep strain can be finished in half a year. Then, the increasing rate of the creep strain gradually decreases and will stabilize about 3 years later. The creep strain measured two years later is 1~4 times of the instantaneous strain. If unloaded at this time, the specimen will recover part of the deformation (i. e. elastic recovery ε'_{ela}), which is less than the instantaneous strain at loading. Another part of strain (i. e. creep recovery ε''_{ela}) can be recovered in about 20 days after unloading. However, most of the strain is unrecoverable, which is called the residual strain ε'_{cr}.

The experiment shows that the creep of concrete is closely related to the stress of concrete (Figure 2-15).

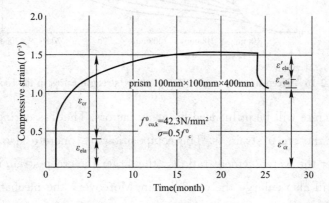

Figure 2-14 Creep of concrete
(Strain and time curve)

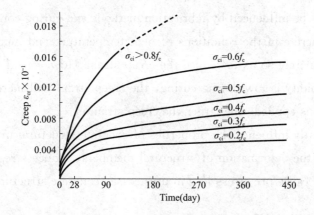

Figure 2-15 The relationship between stress and creep

When stress of concrete is large (larger than $0.5f_c$), creep strain is not proportional to the stress any longer, and comparatively the increase rate of creep is greater than that of stress. This phenomenon is called nonlinear creep. When loading force is too high, creep strain will grow rapidly to no convergence as shown in Figure 2-16. Since the long-term loading can cause the destructive effect to concrete, it is generally agreed that the compressive strength of concrete under sustained loads is just 75%~80% of its short-term counterpart and long-term high strength should be refrained.

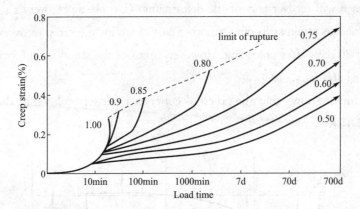

Figure 2-16　Creep-time curve of different stress/strength ratios in log coordinate

Creep of concrete will be influenced by many factors. The less curing time of concrete takes, the larger the creep strain is. The composition of concrete also affects the creep greatly. The larger the water/cement ratio is, the larger the creep strain is. The increase in cement content will also enlarge the creep strain. Moreover, the mechanical properties of aggregates have an apparent influence on the creep of concrete. For example, using maximum sizeof solid aggregates allows an increased modulus of elasticity, and an increased volume ratio of aggregates to concrete helps reduce the creep strain.

The creep may be influenced by fabrication methods and curing conditions of concrete as well. Curing concrete in the conditions of high temperature and humidity can promote the hydration of concrete so as to reduce the creep strain. However, if the temperature is high while the humidity is low during curing, the creep strain will increase. Additionally, the earlier the load is applied, the larger the creep strain is.

Creep will greatly influence the properties of concrete structural members. For example, it can enlarge the deformation of structural members, induce stress redistribution in cross sections, and cause pre-stress loss in prestressed concrete structures.

3. Shrinkage and swelling deformation of concrete

Shrinkage refers to the volume decrease of a concrete member when it loses moisture by evaporation. The opposite phenomenon, swelling, occurs when the volume increases through water absorption. Shrinkage and swelling represent the volume change of concrete specimens during hardening irrespective of the external load. Figure 2-17 illustrates the test of free shrinkage of concrete. Shrinkage grows with time, and shrinkage value of concrete under steam curing is less than the shrinkage value under room temperature curing.

If concrete shrinkage is restrained by surrounding constraints or because of adverse ambient condition, cracks will happen at members' surface.

Here are several factors that affect the degree of drying shrinkage.

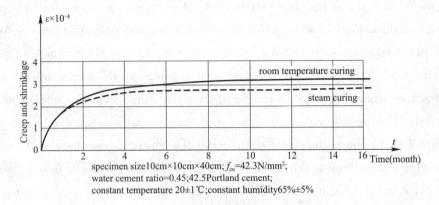

Figure 2-17 Shrinkage of concrete

(1) Type of cement: the higher the cement grade is, the larger the shrinkage strain will be.

(2) Amount of cement: the larger the cement usage or the water/cement ratio is, the larger the shrinkage strain is.

(3) Aggregate property: the larger the modulus of elasticity of aggregate is, the smaller the shrinkage strain is.

(4) Ambient conditions: in the hardening and following service stage of concrete, the larger the ambient moisture is, the smaller the shrinkage strain is. The shrinkage strain will decrease with the increase of curing temperature if the relative humidity is large. And an opposite trend will be observed if the ambient condition is dry.

(5) Quality of construction: the denser vibration condition leads to smaller shrinkage strain.

(6) Service Environment: the shrinkage strain will decrease with the increase of service environment temperature and humidity.

(7) Volume to surface area ratio: larger value of volume to surface area ratio can reduce the shrinkage strain.

2.1.4 Fatigue Performance of Concrete

The strength and deformation of concrete under repeated loading (several loading and unloading cycles) are greatly different from those under monotonic loading. Fatigue failure can happen to concrete under repeated loading.

Specimens sized 100mm × 100mm × 300mm or 150mm × 150mm × 450mm are usually used in fatigue tests of concrete. And the compressive stress at which the concrete specimen finally fails after 2 million (or even more) times of repeated loading is called the fatigue strength of concrete.

Figure 2-18(a) shows the stress-strain curve of a concrete prism subjected to one cycle of loading and unloading, in which OAB presents loading and unloading process. When the stress is decreased to zero after having reached point A, most of the overall strain ε_c corresponding to point A, i. e., ε_e', can be recovered instantaneously during unloading, and a small portion of strain ε_e'' can also be recovered after some time, which is referred to as elastic hysteresis. The unrecovered strain ε_{cr}' is called the residual strain.

Figure 2-18(b) shows the stress-strain curve of a concrete prism subjected to many cycles of loading and unloading. When the loading stress is lower than the fatigue strength of concrete f_c', e. g., σ_1 or σ_2 in Figure 2-18 (b), the stress-strain curve is similar to that in Figure 2-18 (a). But the loop formed by loading and unloading curves in one cycle tends to be closed after several repeated times. However, even if the repeated times are as high as several million, the concrete prism will not be failed by fatigue. If the loading stress exceeds the fatigue strength f_c^f, i. e. σ_3 in Figure 2-18(b), the loading curves that initially convex to the stress axis will gradually become convex to the strain axis after many times of repeated loadings. In the final stage, the loading and unloading curves in one cycle cannot form a closed loop and the slope of the stress-strain curve continuously decreases, indicating the imminent fatigue failure of concrete.

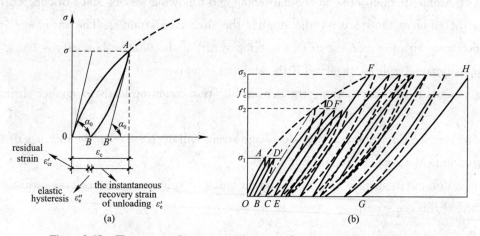

Figure 2-18 The compressive stress-strain curve of concrete under repeated load

(a) curve under one loading cycle; (b) curve under several loading cycles

Concrete fatigue originates from internal defects such as micro-cracks and micro-voids. The stress concentration in concrete under repeated loading causes defects to develop and form macro-cracks, which finally lead to concrete failure. Fatigue failure is brittle, i. e., giving no apparent warning before failure. Cracks are not wide, but deformation is large.

The fatigue strength of concrete is related to the magnitude of stress change under repeated loading. At the same number of repetitions, fatigue strength increases with fatigue stress ratio decreasing. The fatigue stress ratio ρ_c^f is calculated as follows:

$$\rho_c^f = \frac{\sigma_{c,\min}^f}{\sigma_{c,\max}^f} \quad (2\text{-}12)$$

Where $\sigma_{c,\min}^f$ and $\sigma_{c,\max}^f$ represent the minimum and maximum stresses of concrete in the same fiber, respectively.

Code for Design of Concrete Structures GB 50010 stipulates that the design value of concrete fatigue strength is determined by the design strength f_c or f_t multiplied by corresponding corrector factors γ_p.

2.2 Mechanical Properties of Steel Bars

2.2.1 Type of Steel Bars

The steel type used in reinforced concrete members can be classified as flexible reinforcement and stiffness reinforcement.

1. Flexible steel bars

All the linear steel bars can be named flexible reinforcement, and steel bars can be classified as plain bars and deformed bars according to their surface profile. Deformed bars are the bars with longitudinal and transverse ribs rolled into surfaces. The ribs are in the shape of spiral, chevron or crescent (Figure 2-19). The cross-sectional area of a deformed bar varies with its length, so the diameter of the deformed bar is a nominal dimension, which is an equivalent diameter same with the plain bar of identical weight. Crescent stiffness bar has been used widely in China because of its weak stress concentration.

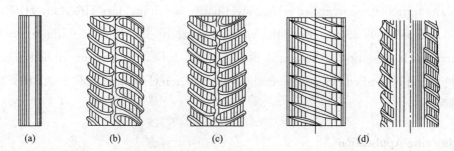

Figure 2-19 Shape of bars

(a) Plain round bar; (b) Spiral reinforcement; (c) Herringbone reinforcement; (d) Crescent bar

Small diameter steel bars (e.g. $d<5mm$) are also called steel wires, whose surface is generally smooth. Sometimes, people also make indentations into the surface to improve the bond.

2. Stiffened steel bars

Stiffened steel bars include shape steels and the skeletons fabricated by welding several pieces of shape steels together. Due to the large stiffness, stiffness reinforcement can be used in construction forms to bear the self-weight of structure and construction loads. This can facilitate shuttering work and speed up construction process. Additionally, structural members reinforced by stiffness reinforcement possess higher loading capacity than those reinforced by steel bars. And stiffness reinforcement is widely used in beams, columns, shear walls and tube structure of high buildings.

2.2.2 Domestic Ordinary Steel Bars

This chapter will mainly discuss the ordinary steel bars, and the prestressed reinforcement will be mentioned in Chapter 9.

According to *Code for Design of Concrete Structures* GB 50010, the ordinary steel bars used in China for reinforced concrete are hot-rolled steel bars. Hot-rolled steel bars are the mild steel rolled at high temperature. The material can be low carbon steel bars or ordinary low-alloy steel bars. Its stress-strain curve has apparent yield point and yield plateau, and the steel shows area reduction at the weakest cross section and the elongation is large.

1. Strength grade and grade of steel bars

The grade of strength classified by yield strength of ordinary steel bar in China can be designated as 300MPa, 335MPa, 400MPa and 500MPa.

There are 8 grades for ordinary steel bars in China. HPB is the abbreviation of hot-rolled plain steel as the symbol of Φ, of which the standard yield strength is 300MPa. HRB is the short term of the hot-rolled ribbed steel bars, which includes HRB335, HRB400 and HRB500, respectively representing the yield strength of 335MPa, 400MPa and 500MPa. For further classification, there are also remained heat treatment ribbed steel bars (RRB400) and hot-rolled ribbed fine-grained steel bars including HRBF335, HRBF400 and HRBF500.

2. Engineering Application

Code for Design of Concrete Structures GB 50010 encourages the wide application of high strength steel in construction. Therefore, the main steel bars used in beams and columns are mainly HRB400 (⏀) and HRB500 (⏀). The design strength of material equals to the standard strength divided by the coefficient. The coefficient of HRB400 is 1.1, so its design strength equals to 360N/mm^2.

Stirrups are always chosen from HRB400, HRBF400, HRB335 and HPB300.

Hot-rolled plain bars can be used for longitudinal reinforcement, but considering its lower strength, HPB300 is mostly served as stirrup. HRB 500 and HRBF 500 steel bars can only be used in spiral hooping when functioning as stirrups.

HRBF500, HRB500 and RRB400 steel bars all cannot carry fatigue loads, and HRB400 is a better alternative.

The grades of steel are also called Grade Ⅰ, Ⅱ, Ⅲ and Ⅳ reinforcement, but this name form is forbidden in official files and drawings.

2.2.3 Strength and Deformation of Steel Bars

Typical stress-strain curves of steel bars used in reinforced concrete structures are obtained from monotonic tension test, in which the loads are monotonically applied until the failure of specimens in a short time.

From monotonic tension tests, researcher can evaluate the strength and deformation of steel bars. Figure 2-20 and Figure 2-21 show two stress-strain curves of steel bars with apparent differences.

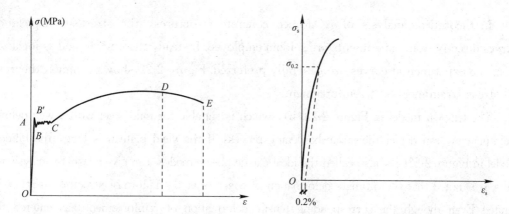

Figure 2-20　The stress-strain curve of steel bar with obvious flow amplitude

Figure 2-21　The stress-strain curve of steel bar without apparent flow

For hot-rolled low-carbon steel and hot-rolled low-alloy steel, the stress-strain curve in Figure 2-20 is recorded. The curve exhibits an initial linear elastic portion. The stress corresponding to point A is called the proportional limit. In segment AB, the strain increases a little bit faster than the stress, but it is not very obvious in the curve. After point B, the strain grows rapidly with little or no increase in the corresponding stress. The curve extends nearly horizontally to point C. Segment BC is called the yield plateau. After point C, the stress increases again with the strain until to the point D. The stress corresponding to the highest point D is the ultimate strength of steel bars. Segment CD is the strain-hardening stage. After point D, the strain increases rapidly accompanied by the area reduction of the weakest cross section, and finally fracture occurs at point E.

For high-carbon steel, Figure 2-21 shows the stress-strain curve and no apparent yield plateau can be observed in the curve. Generally, the stress $\sigma_{p0.2}$ corresponding to the residual strain at 0.2% is taken as the yield strength. It has been stipulated in Chinese metallurgical standards that the yield strength $\sigma_{p0.2}$ of reinforcement should not be less than 85% of the ultimate tensile strength σ_B.

Steel bars should not only have enough strength, but also behave with plastic shrinkage. Originally, plastic shrinkage of steel bars are judged by uniform elongation and cold bending property. The ratio of deformation and original length is defined as elongation. The higher the elongation is, the better the plastic shrinkage will be. Cold bending is measured by buckling the steel bar of diameter d around the steel roller with diameter D to a specified angle with no cracks or fractures. And the larger the diameter of the steel roller is, the better the plastic shrinkage will be.

2.2.4 Constitutive Model of Steel Bars

In theoretical analysis of reinforced concrete structures, the stress-strain relation curves directly from experiments are seldom employed. Instead, theoretical models idealized from the experimental curves are generally preferred. Figure 2-22 shows common theoretical stress-strain models of reinforcement.

The trilinear model in Figure 2-22(b), which is suitable for mild steel with a well-defined yield plateau, can depict the strain-hardening process. If the yield plateau is long, the bilinear model in Figure 2-22(a), also called the ideal elastic-plastic model, can give adequate analysis results. It is noted that the ultimate deformation of concrete at the failure of structure members is limited. Even though the corresponding tensile deformation of reinforcement has entered the strain-hardening stage, the hardening effect can be ignored due to the limited extent. Therefore, in practical engineering, the elastic-perfectly plastic model is commonly employed for ordinary steel bars in theoretical analysis, which can be formulated as:

$$\text{if } \varepsilon_s \leqslant \varepsilon_y, \; \sigma_s = E_s \varepsilon_s \left(E_s = \frac{f_y}{\varepsilon_y} \right) \tag{2-13}$$

$$\text{if } \varepsilon_y \leqslant \varepsilon_s \leqslant \varepsilon_{s,h} (\sigma_s = f_y) \tag{2-14}$$

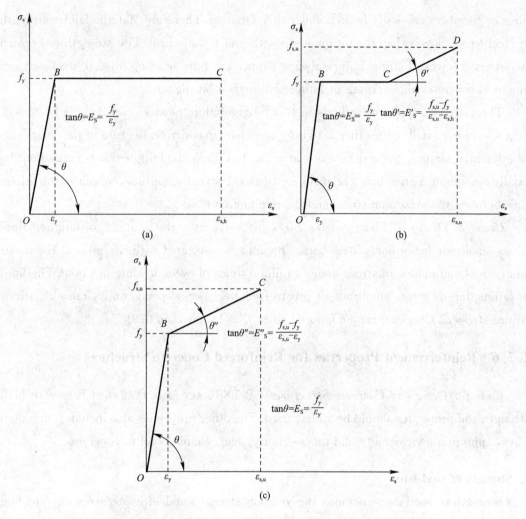

Figure 2-22 The mathematical models of the stress-strain curve
(a) Dual linear; (b) Triple line; (c) Double slash

2.2.5 Fatigue Failure of Steel Bars

When a steel bar is subjected to periodic loading, the steel bar will fail after a certain number of times of loading and unloading between the minimum stress σ_{min}^f and the maximum stress σ_{max}^f, even though the maximum stress is lower than the strength under monotonic loading. This is called fatigue failure. In engineering applications, fatigue failure may happen to reinforced concrete members under repeated loading, such as crane beams, bridge deck and sleepers.

The reason for fatigue failure of steel bars is generally due to the defects in inner and outer space of steel, where it is easy to cause stress concentration. Grains of steel will slide when stress is high, and the increase of repeated loading will lead to wider crackles. Structures or members are easily broken under this situation. Therefore, fatigue failure strength of steel bars is lower than the ultimate strength under static load. The strength of original steel bars is lowest. Fatigue failure always happens at pure bending area. If steel bars are buried in concrete, the strength of fatigue failure is a bit higher.

There are two types of methods to test fatigue failure of steel bars. One is by tensioning a steel bar axially, the other is to bury steel bars in concrete. Because of the complicated influential factors, the test data is scattering. In China, the fatigue test is carried out by axially tensioning a steel bar. The main factor is the stress amplitude, which is the difference between the maximum stress and the minimum stress.

Code for Design of Concrete Structures GB 50010 rules the limit Δf_y^f of fatigue failure stress amplitude for ordinary steel bars. The limit is associated with the ratio of the maximum stress and minimum stress under 2 million times of cyclic loading in China. The limit of fatigue failure stress amplitude of prestressed steel bars depends on its ratio of fatigue failure stress ρ_p^f. Checking can be ignored when ρ_p^f is bigger than 0.9.

2.2.6 Reinforcement Properties for Reinforced Concrete Structures

Code for Design of Concrete Structures GB 50010 suggests that steel bars with high strength and properties should be widely used. The other properties also include: high ductility, appropriate weldability and processibility, and reliable bond to concrete.

1. Strength of steel bars

Strength of steel bars includes the yielding strength and ultimate strength. Yielding strength is one of the main indexes in designing and calculation. Using high strength steel bars can save the costs and amount of steel.

2. Ductility of steel bars

Steel bars with good ductility can ensure enough deformations before failure. And high property of cold bending is also necessary at the same time. Elongation and cold bending property are the main indexes when checking steel bars.

3. Appropriate weldability of steel bars

Appropriate weldability is one index of connection property. Good appropriate weldability requires no crackle on the surface.

4. Property of Mechanical Connection

It is a better choice to take mechanical connections between steel bars, which is widely used in China.

5. Adaptability of construction

Steel bars should be machined and installed easily in construction.

6. Cohesive force between steel bars and concrete

In order to have a good working cooperation between steel bars and concrete, enough cohesive force is required. And the surface shape of steel bars is a significant factor to the cohesive force.

In low temperature regions, performance relevant to low temperature are also needed to be considered.

2.3 Bonding Between Steel Bars and Concrete

2.3.1 Importance of Bonding

Bonding behavior is the interaction between steel bars and the surrounding concrete. It can be classified into two types, one is in length of steel bar, and the other is anchorage at ends. And bonding between steel bars and concrete is the foundation to co-work of steel bars and concrete.

Bonding can be clarified by Figure 2-23. Because of the different load properties, bonding between steel bars and concrete can be classified as local bond stress and bond stress at ends. Local bond stress occurs between two cracked section, where the concrete is in tension. The stress is distributed non-uniformly. The loss of local bond stress will cause the decrease of strength and development of cracks. At ends, a certain length of steel bars should be extended, which is called the anchorage length. To protect the safety of members, extending enough length is required to accumulate enough bonding strength. And hooks, buckling and welding short steel bars are taken to improve the bonding strength. For tension plain steel bars, hooks must be set at ends.

2.3.2 Components of Bonding Strength

Plain steel bars have different mechanism from deformed steel bars.

Bonding strength between plain steel bars and concrete is composed of three parts as follows:

(1) The cementing force between steel bars and the surface of concrete. This kind of force is generated by cement paste and the oxide layer. Cementing force is not obvious and will vanish if there is slide at interface.

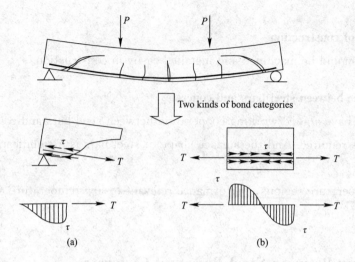

Figure 2-23 A schematic diagram of bond stress between reinforcement and concrete
(a) Anchorage bond stress; (b) Local bond stress between cracks

(2) The friction force between steel bars and concrete. Concrete will shrink as well as when it begins to freeze, leading compression force around the side of steel bars. The higher the compression force and the roughness of the surface are, the larger the friction force will be.

(3) The mechanical force between steel bars and concrete. Nevertheless, plain steel bars only rely on its surface drawback.

For deformed steel bars, originated from steel bars embedded into concrete, the mechanical force is the main bonding strength. Mechanical force strengthens the interaction between steel bars and concrete, which largely improves the bonding strength. Figure 2-24 shows the principle of components of bonding strength.

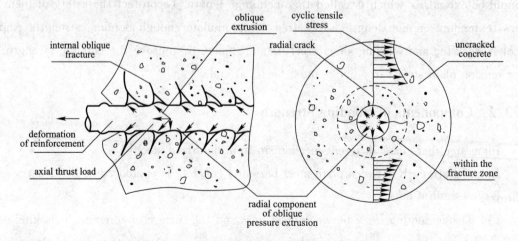

Figure 2-24 The inside crack of the concrete around the rib bar

It can be concluded that the main difference between plain steel bars and deformed steel bars is that bonding strength of deformed steel bars is mainly originated from mechanical force.

2.3.3 Bond Stress-Slip Relationship

The bonding property of steel bars and concrete is mainly reflected on their bonding stress τ and relative displacement s.

Figure 2-25(a) shows the bond stress-slip relationship curves of plain steel bar tension test. It can be seen that bonding stress of plain steel bars with no rust is smaller. At the peak point, the friction force begins to fall down and displacement of slide continuously increases. Corrosion plays a significant role for bonding stress.

Figure 2-25(b) shows the bond stress-slip relationship curve of deformed steel bars tension test. At first, ribs of steel bars exert oblique pressure on concrete, because of the large rigidity, displacement of slide is short. The relationship between bonding stress and relative displacement resembles a line; with the oblique pressure rising, crackles happen at inner area, reducing the rigidity and increasing the displacement of slide, which leads to the slope reduction; When the oblique pressure rises to an extent, concrete is broken, and the new sliding surface generates. At last, crackles reach the surface of specimen, and the bonding stress reaches the peak point. The ultimate sliding length is about 0.35~0.45mm.

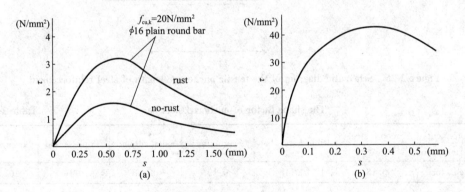

Figure 2-25 τ-s curve

(a)τ-s curve of plain round bar; (b) τ-s curve of ribbed bar

2.3.4 Anchorage of Steel Bars

1. Basic length of anchorage

The *Code for Design of Concrete Structures* GB 50010 stipulates the anchorage length of tension steel bars l_{ab}, which is the basic length of anchorage.

The principle of the anchorage length of tension steel bars is given as Figure 2-26. The diameter of the steel bar in the specimen is d. When its stress equals to design tension strength, the pressure is a quarter of $f_y \pi d^2$. Supposing that the average bonding strength equals to τ, the total bonding strength is caculated $\tau \pi d^2 l_{ab}$, τ is supposed to calculate as f_t divided by 4α. According to the equilibrium of forces,

$$l_{ab} = \alpha \frac{f_y}{f_t} d \qquad (2\text{-}15)$$

Where l_{ab}——the basic anchorage length of tension steel bars;

f_y——the design tension strength of steel bars;

f_t——the design tension strength of concrete;

d——the diameter of steel bars;

α——factor of anchorage steel bars, which can be obtained from Table 2-1.

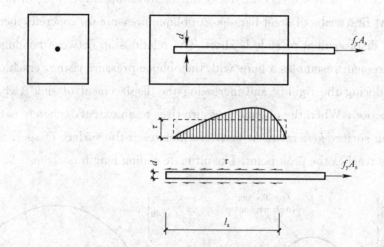

Figure 2-26 Schematic diagram of the tensile anchorage length of steel reinforcement

The shape factor of anchored steel bar   Table 2-1

Reinforcement type	Smooth steel bars	Ribbed steel bar	Scoring bar	Spiral rib wire	Triple strand	Seven strand
Shape factor	0.16	0.14	0.19	0.13	0.16	0.17

It can be easily found that basic anchorage length is the integral multiple of diameter of steel bars.

2. Anchorage of tension steel bars

(1) Anchorage length l_a of tension steel bars

Anchorage length of tension steel bars in construction should consider different conditions, with no less than 200mm.

$$l_a = \xi_a l_{ab} \qquad (2\text{-}16)$$

Where l_a——anchorage length of tension steel bars;

ξ_a——factor of anchorage length, the value is taken according to the following situations:

1) The value is 1.10 if nominal diameter of ribbed steel bars is larger than 25mm;

2) The value for epoxy resin coated steel bars is 1.25;

3) The value for steel bars to be disturbed easily is 1.10;

4) The correction factor is the design calculation area divided by actual reinforcement area when actual reinforcement area is bigger than design calculation area. However, correction factor should be ignored when structures under dynamic load or earthquake fortification;

5) The value of correction factor is 0.80 when the thickness of anchorage steel bars' protective layer is more than $3d$; and value is selected as 0.70 when the thickness of anchorage steel bars' protective layer is $5d$. Values are chosen by interpolation when the thickness of anchorage steel bars protective layer is between $3d$ and $5d$. d is the diameter of anchorage steel bars;

6) If correction factors are not merely influenced by one factor, then all the factors should be combined together, but the total factor should be no less than 0.6; and value of prestressed steel bars is taken 1.0.

(2) Transverse structural reinforcement in anchorage area

The diameter of transverse structural reinforcement should not be less than $0.25d$ at anchorage area when the thickness of anchorage area is no more than $5d$.

(3) The manners of anchorage

When the end of ordinary tension steel bars are conducting mechanical anchorage or bending anchorage, the anchorage length including the hooks and the end is $0.6l_{ab}$. The measures and technical demand are as follows (Figure 2-27):

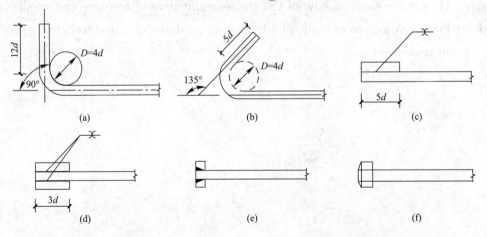

Figure 2-27 The form and technical requirements of hook and mechanical anchorage

(a) 90° hook; (b) 135° hook; (c) One side welding; (d) Two sides welding;
(e) Perforated plug anchor plate; (f) Bolt anchor head

3. Anchorage of compression steel bars

The anchorage length of compression steel bars of concrete should not be less than 70% of tension anchorage length when the compression strength is considered fully.

The hooks and one side welding should not be adopted in compression steel bars.

The transverse structural reinforcement in the anchorage length area of compression steel is the same with tension steel.

Anchorage of stressed steel bars at supports in beams or floor and connection of steels will be discussed in Chapter 4.

Questions

2-1   How is the reinforcement classified?

2-2   What is the difference between the stress-strain curves of hard steel and mild steel? How is their yield strength determined?

2-3   What mathematical models are used to simulate the stress-strain relationship of reinforcement? How are their applicable conditions described?

2-4   What is the influence of cold drawing and cold stretching on mechanical properties of steel?

2-5   What are the requirements on properties of steel in reinforced concrete structures?

2-6   How are the cube strength, axial compressive strength, and the axial tensile strength of concrete determined?

2-7   How is the characteristic of the stress-strain curve of concrete under axial compression? Please provide an example of a frequently used mathematical model to define the stress-strain relationship.

Chapter 3
Load-Carrying Capacity on Normal Section for Flexural Members

3.1 General Structure of Beams and Slabs

Flexural members have various section shapes, e. g. , rectangular section, T-shaped section, I-shaped section, channel section, box section and inverted L section of symmetry or asymmetry as shown in Figure 3-1.

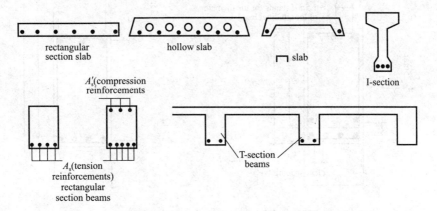

Figure 3-1 Section types of beams and slabs

The height-to-width ratio h/b is 2~3.5 for a rectangular section, and 2.5~4 for a T section. The beam width b usually takes 100mm、120mm、150mm、(180mm)、200mm、(220mm)、250mm and 300mm and over 300mm, thereafter increasing by a 50mm increment. The number in the parentheses is only used for wooden members. The beam height h usually takes 250mm、300mm、350mm、750mm、800mm、900mm、1000mm, increasing by 50mm increment under 800mm and 100mm increment above 800mm.

The thickness of solid reinforced concrete slabs often increases by a 10mm increment. The minimum thickness is 60mm for roof panels and residential floor slabs, 70mm for industrial floor slabs, 120mm for trough plates in railway bridges, 100mm for traffic lane panels, and 80 mm for sidewalk slabs.

The concrete beams and slabs usually use the concrete grade as C25 or C30. In order to prevent excessive shrinkage of concrete, the concrete grade should not be greater than C40.

The longitudinal steel bars in beams use HRB400 and HRB500. The diameter of longitudinal steel bars in beams usually takes 12mm, 14mm, 16mm, 18mm, 20mm, 22mm and 25mm. When the beam height h is greater than 300mm, the diameter should not be less than 8mm.

The minimum number of longitudinal bars is 2 (1 is acceptable if $b<100$mm). In order for coarse aggregates to pass through a cage of steel bars freely so that the concrete can be appropriately compacted, to ensure a good bond between steel bars and concrete and to provide enough protection to reinforcement, the clear spacing between longitudinal bars must satisfy the requirements illustrated in Figure 3-2.

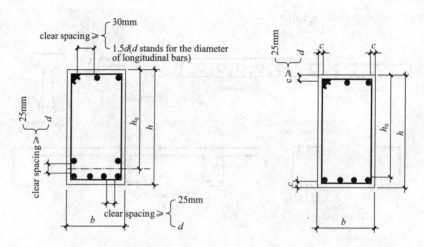

Figure 3-2　A rectangular section beam

The diameter of auxiliary steel bars is usually 8mm when the span of beam is less than 4m, 10mm when the span is between 4m and 6m, and 12mm when the span is greater than 6m.

The grade of steel bars for stirrups are HRB400 and HRB335, and the diameter of longitudinal bars is normally $6\sim12$mm in solid slabs. In order to facilitate the pouring of concrete to ensure the density of the concrete around the rebar, spacing of bars should not be too close; meanwhile, in order to share the internal force effectively, it should not be too large. Reinforced spacing is generally $70\sim200$mm; when the thickness of plate is $h\leqslant150$mm, it should not be greater than 200mm; when the thickness is $h>150$mm, it should not be more than $1.5h$ and 250mm. The distribution bars adopt HRB400 and

HRB335 with a diameter usually ranging from 6mm to 8mm. The section area of the distribution bar on the unit width should not be less than 15% of the longitudinal bar in the unit width and the reinforcement ratio should not be greater than 0.15%. The bar spacing is shown in Figure 3-3.

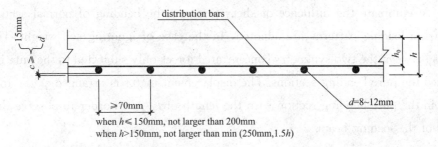

Figure 3-3 A rectangular section slab

Define $\rho = A_s/bh_0$ as the reinforcement ratio of longitudinal bars in a beam, where h_0 (effective depth) is the distance from the edge of compression fiber to the centroid of the steel bars, b is the section width and A_s is the area of longitudinal steel bars. The bending behaviors of beams with different reinforcement ratios will be studied accordingly.

$$\rho = A_s/bh_0(\%) \tag{3-1}$$

Different requirements are specified in various design codes for minimum concrete cover c, which is generally defined as the shortest distance between the outer surfaces of longitudinal bars and the nearby surface of a member.

Similar to axially loaded members, flexural members should be constructed in accordance with appropriate detailing requirements to ease construction and achieve a good bond between steels and concrete. Appropriate concrete cover also helps protect steel bars from corrosion and fire disasters.

Figure 3-2 illustrates the cover thickness for beams and slabs in a normal indoor environment. An extra 5mm should be added to the minimum cover thickness if the concrete grade is lower than C20.

3.2 The Flexural Property of Normal Section for Flexural Members

3.2.1 Three Typical Stages for Beams with Under-Reinforced Beam

The following sections will cover bending failure modes and longitudinal tensile reinforcement ratio of normal section for flexural members. When the longitudinal tensile bars can lead to the bending failure form of the normal section as the type of ductile damage, it is called the under-reinforced beam.

1. Test setup

Figure 3-4 shows a typical test setup for the investigation of mechanical behavior of a simply supported and singly reinforced concrete beam. The concrete strength of the design is C25. To eliminate the influence of shear force on the bending of normal section, two symmetrical loading patterns are adpoted, in the case of ignoring self-weight. The cross sections between the two symmetry concentrated forces only subjected to the pure bending, are called the pure bending sections. The displacement meter is arranged at the top of the section in the pure bending section with the length of $l_0/3$, in order to observe the whole process of the loading beam.

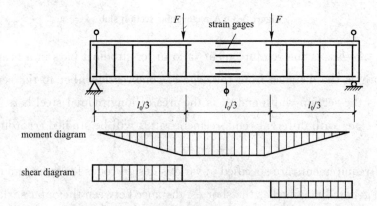

Figure 3-4 Reinforced concrete beam under bending

The load is applied by stage, starting from zero until the beam is bent. In order to study the whole process of the force during the loading process, the longitudinal deformation of the measuring points on the high side of the beam is measured in the pure bending section. Therefore, before pouring concrete, the strain near the middle sections will be measured. Because the measuring instrument always has a scale distance, the measured value indicates the average strain value in the range of the subscale. Furthermore, the micrometers are installed at the mid-span and the support in order to measure the deflection f, and sometimes the inclinometer is needed to measure the intersection angle of the beams.

In the pure bending section, the bending moment will rotate the normal section. In the unit length of the beam, the angle of the rotation in the normal section is called the sectional curvature, which represents the bending deformation of the normal section, the unit is 1/mm.

Figure 3-5 shows the moment-curvature relationship curve of reinforced concrete from the China Academy of Building Research. The graphic contains the experiment value of mid-span sectional curvature of beams φ^0 on the horizontal axis and experiment value of mid-span bending moment of the beams M^0 on the vertical axis. The superscript "0" represents the experimental value.

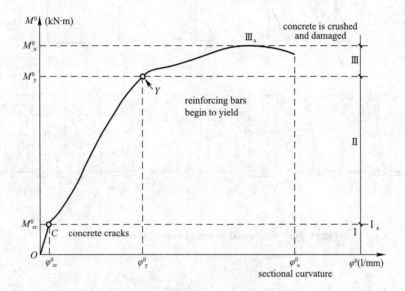

Figure 3-5 Load against deflection of reinforced concrete beam

There are two obvious turning points C and Y on the curve, which divide the curve into three stages: elastic stage, service stage and failure stage.

2. The three stages during the loading process

Stage I: Elastic stage

At first loading, because of the small curved distance, the fiber strains measured along the beam are small, too, which are linear along the height of the cross section. The stress of the beam is similar to that of the homogeneous elastomer beam, and the stress is proportional to the strain. The stress distribution diagram of concrete in compression zone and tensile zone is triangular (Figure 3-6).

Due to the weak tensile strength, with the increase of the moment, the concrete at the edge of the tensile zone firstly shows the plastic characteristics that the strain is growing faster than stress. The stress distribution diagram in the tensile zone begins to deviate from the line and gradually becomes a curve. The bending moment continues to increase, and the area of the curved part for the stress distribution diagram in the tensile region expands to the neutral axis constantly.

In general, characters in the first stage are: 1) The concrete does not crack; 2) The concrete stress graphic of compressive zone is linear while concrete stress graphic of tensile zone undergoes from a straight line at the early stage to a curve subsequently; 3) Bending moment and cross section curvature basically show linear relationships.

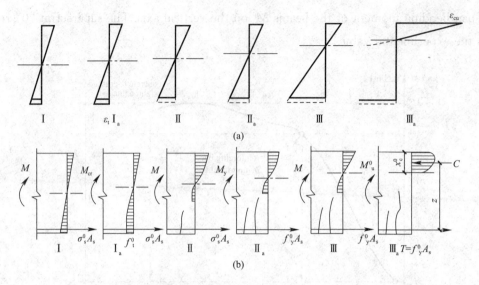

Figure 3-6 Stress and strain distribution across section
(a) Strain distribution; (b) Stress distribution

Stage I can be used as the basis for calculating the crack resistance of flexural members.

Stage II: Service stage

For $M^0 = M_{cr}^0$, the first crack will appear in the most vulnerable section of the pure bending when the tensile strain of the edge fiber reaches the experiment value of the concrete limit strain. Once the member is cracked, the beam will turn the stage I into stage II.

After cracking, the tension force originally undertaken by the concrete will transfer to the steel bar, which makes the stress of reinforcement suddenly increase, and the deflection of the beam and the curvature of the cross section suddenly grow as well. The neutral axis of the fracture section will also move up. The concrete still supports a small amount of tension in the area under the neutralization axis but the effect can be ignored, where the tension of the tensile force is mainly borne by the steel bar.

With the increase of bending moment, the measured values of the concrete compressive strain and tensile strain are both constantly increased. Although across several cracks of larger values, the measured strain is still consistent with linear relationship along the height (Figure 3-6).

With the bending moment increasing constantly, the curvature of the section increases, and the cracks grow wider and wider. Due to the increasing strain of the compressive zone, the concrete strain is growing faster than the stress, and the plastic characters become more and more obvious, as shown a curve in the compression zone (Figure 3-6).

The stage II describes the occurrence of the fracture in the section. In this section the beam is working with cracks. The stress characters are: 1) the steel bars have not yielded even after the concrete has been crushed; 2) The concrete in the compression zone has plastic deformation, but does not reach the limit state, and the compression diagram is only within the curve of the ascending segment; 3) The compressive stress of the concrete in the compression zone changes from linear distribution to nonlinear distribution with the increasing load increments.

Stage II is normal condition, which can be used as the basis of checking deformation and crack width in the service condition.

Stage III: Failure stage

After the yielding of tensile bars, a small load increment will produce large deformation of the beam. Cracks propagate further and become wider, which pushes the neutral axis upward, and the nonlinear distribution of compressive stress of concrete becomes more apparent (see Figure 3-6).

When compression strain of the edge fiber reaches or close to the experimental value of the limit state, it is the symbol before failure, approximate as 0.003~0.005.

The damage of cross section begins at the yield of the longitudinal tensile steel bar and ends up with the concrete crack at the edge of the compression zone. The characters are: 1) The longitudinal tensile steels have been yielded, and the tension remains constant. Most of the concrete in the crack section is out of work while the concrete curve in the compression zone develops more complete. 2) The force arm is increased due to the total concrete compression force in the compression zone, and the bending moment has increased slightly. 3) When the compression strain reaches the experimental value of the limit state, the concrete is crushed, and the cross sections are damaged. 4) Bending moment-curvature relation is an approximate horizontal curve.

This stage can be used as the basis for calculating the bending capacity of the normal section.

3. The characteristics of the whole process

In conclusion, the test beam has the following characteristics from loading to destruction:

(1) In stage I, the growth of the cross section curvature or deflection of beam is slow; In stage II, due to the beam working with cracks, the growth is faster than before; In stage III, the curvature of the section and the deflection of the beam are greatly increased due to the yield of reinforcement.

(2) With the increase of bending moment, the neutral axis keeps moving upward, and the height of the compression zone x_c^0 gradually reduces. The compression strain on the concrete edge and the tensile strain of reinforcement also increase with the growth of ben-

ding moment, but the average strain is still consistent with the plane section hypothesis, that is to say the tensile stress of concrete is roughly corresponding to the single axis stress curve of the concrete; In stage Ⅲ, the graphics of the compressive stress are roughly corresponding to uniaxial compressive stress curve.

(3) As shown in Figure 3-7, the reinforcement stress grows slowly in stage Ⅰ; When $M^0 = M_c^0$, the reinforcement stress mutates, and the reinforcement stress in stage Ⅱ grows faster than that in stage Ⅰ; When $M^0 = M_y^0$, reinforcement stress reaches yield strength.

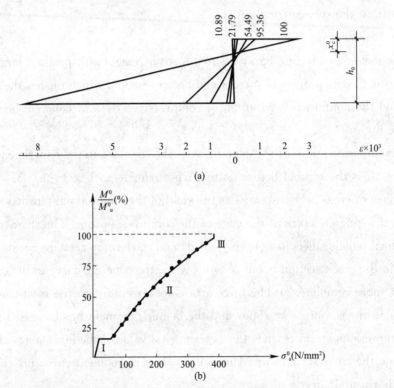

Figure 3-7　The change of strain along the section height and the measured results of reinforcement stress

The experiments and researches have shown that the stress of reinforced concrete structure and components can also be divided into three stress stages. Therefore, the three stress stages are the basic properties of reinforced concrete structures, thus it is very important to know the three stress stages correctly.

To be attentioned, the following three viewpoints are wrong: 1) The stage Ⅰ is the elastic stage; 2) The concrete will crack when it reaches the tensile strength; 3) The concrete will crack when it reaches the crushing strength.

3.2.2　Failure Modes

The structure, components and sections usually have two types of damages, namely brittle failure and ductility failure. Before the destruction, the deformation is very small, and there is no obvious signal to destruction, which is defined as the brittle failure; Before the destruction, the deformation is large with obvious signs to destruction, which belongs to the type of ductile damage.

The experiment shows that due to the difference of the longitudinal tensile reinforcement ratio, the bending failure of the normal section is divided into three kinds, namely failure mode of under reinforced beam, failure mode of over reinforced beam and failure mode of lightly reinforced beam, as shown in Figure 3-8. Figure 3-8(a), (b) and (c) respectively present the under reinforced beam, over reinforced beam and lightly reinforced beam.

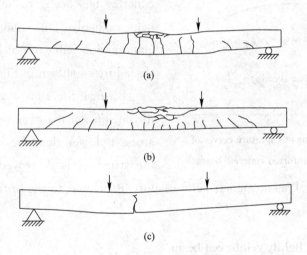

Figure 3-8 Failure modes of reinforced concrete beams

(a)Under reinforced beam; (b)Over reinforced beam; (c)Lightly reinforced beam

1. Failure modes of under reinforced beam.

When $\rho_{min}h/h_0 \leqslant \rho \leqslant \rho_b$, the concrete beams belong to the under reinforced beams. The characteristic is that the failure of this type starts from the yielding of longitudinal steel bars and ends at the crushing of concrete. This kind of failure is called tension failure and is associated with ductile failure. ρ_{min} and ρ_b are the minimum reinforcement ratio and critical reinforcement ratio respectively under this type.

At the beginning when the tensile reinforcement reaches the yield strength, the strain of the edge fiber in the compression zone is less than that in ultimate state. During the process of the reinforcement yielding to the edge concrete cracking, the reinforcement will undergo a large plastic deformation. Then the crack is rapidly developed and the deflection of the beam is greatly increased, which gives a clear warning before destruction (Figure 3-8).

From the Figure 3-5, the increment of moment M_y^0 to M_u^0 is small, while the curvature increment of the cross section φ_y^0 to φ_u^0 is large. This means when the bending moment exceeds M_y^0, there is a great deformation with little bearing capacity changes. In other words, the beam has a good ductility. In chapter 8 the ductility will be explained in detail which is an important indicator to measure the structure capacity or post-deformation of the cross section.

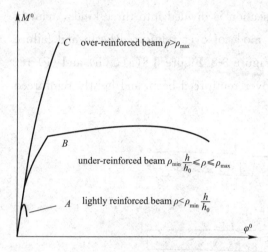

Figure 3-9 The moment-curvature curves of different types of reinforced concrete beams

2. Failure modes of over reinforced beam

When $\rho > \rho_b$, the failure of over reinforced beam occurs. The characteristic is that the steel bars have not yielded even after the concrete has been crushed. The failure happens suddenly. This kind of failure is called compression failure, which is also called brittle failure as shown in Figure 3-9.

Because there are excessive tensile reinforcements on the over reinforced beam, the stress is lower than the yield strength when the structure is destroyed, leading to negative economics as well as no signals before failure. It belongs to the brittle failure, which is not allowed generally.

3. Failure modes of lightly reinforced beam.

When $\rho < \rho_{min} \cdot h/h_0$, the failure of lightly reinforced beam occurs. When the lightly reinforced beam is damaged, the limit bending moment M_u^0 is less than the cracking moment M_{cr}^0. Therefore, the fracture of the damaged concrete also belongs to brittle fracture. The higher the reinforcement ratio is, the greater the difference between $M_{cr0} - M_{u0}$ is. When $M_{cr}^0 - M_u^0 = 0$, it is the limit of the beam in principle.

The crack rapidly propagates to the beam top, and the beam fails in a brittle way. The compressive strength of concrete has not been fully developed.

3.2.3 Balanced Failure and Balanced Reinforcement Ratio

Between tension failure and compression failure, a balanced state exists as balanced failure, when steel yields and concrete crushes simultaneously. The longitudinal reinforcement ratio of beams under balanced failure is called the balanced reinforcement ratio (ρ_b), which is the quantitative index to distinguish tension failure and compression failure. This index is also the maximum reinforcement ratio allowed for under reinforced beams. When $\rho < \rho_b$, destruction begins with the yield of reinforcement; When $\rho > \rho_b$, destruction begins in the edge of compression zone; When $\rho = \rho_b$, the tensile reinforcement stress reaches the yield strength and the concrete crushes in compression zone at the same time. Balanced damage belongs to ductile damage. Therefore, for ductile failure, the reinforcement of the beam should be satisfied as $\rho_{min} \cdot h/h_0 \leqslant \rho \leqslant \rho_b$.

3.3 Calculation Principles for Load-Carrying Capacity of Normal Sections

3.3.1 Basic Assumption

Based on *Code for Design of Concrete Structures* GB 50010, the cross-section bearing capacity of various concrete bending members should be calculated according to the following five basic assumptions:

1. The sections remain plane during the whole process;
2. Concrete is assumed not to sustain any tension;
3. Compressive stress-strain relationship of concrete is taken as the following rules:

When $\varepsilon_c \leqslant \varepsilon_0$

$$\sigma_c = f_c \left[1 - \left(1 - \frac{\varepsilon_c}{\varepsilon_0} \right)^n \right] \tag{3-2}$$

When $\varepsilon_0 < \varepsilon_c \leqslant \varepsilon_{cu}$

$$\sigma_c = f_c \tag{3-3}$$

The parameters are obtained as follows, and $f_{cu,k}$ is the characteristic value of cube compressive strength:

$$n = 2 - \frac{1}{60}(f_{cu,k} - 50) \leqslant 2.0 \tag{3-4}$$

$$\varepsilon_0 = 0.002 + 0.5 \times (f_{cu,k} - 50) \times 10^{-5} \geqslant 0.002 \tag{3-5}$$

$$\varepsilon_{cu} = 0.0033 - (f_{cu,k} - 50) \times 10^{-5} \leqslant 0.0033 \tag{3-6}$$

4. The ultimate strain of longitudinal tensile steel bars is taken as 0.01;
5. The stress of longitudinal reinforcement is product of strain of reinforcement and elastic modulus, and it should meet the following requirements:

$$-f'_y \leqslant \sigma_{si} \leqslant f_y \tag{3-7}$$

σ_{si} is the stress of the steel bar in the layer i, the positive value represents tensile stress while negative value represents compression value.

3.3.2 Equivalent Rectangular Stress Block

In order to simplify the calculation, it is possible to take the equivalent rectangular stress block to replace the theoretical stress pattern of the concrete in the compression zone.

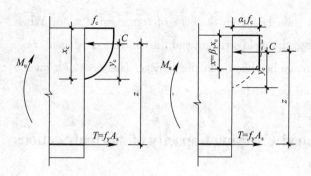

Figure 3-10 Equivalent stress block

As shown in Figure 3-10, the substitutable principles are:

1. The resultant force C of concrete compressive stress should be equal;

2. The centroid position in each graph should be constant.

The equivalent compressive stress of the concrete and the height are defined as $\alpha_1 f_c$ and x, respectively.

$$\left.\begin{array}{l}\alpha_1 f_c bx = k_1 f_c bx_c \\ x = 2(x_c - y_c),\ x = 2(1-k_2)x_c\end{array}\right\} \quad (3-8)$$

Take $\beta_1 = \dfrac{x}{x_c} = 2(1-k_2)$, $\alpha_1 = \dfrac{k_1}{\beta_1} = \dfrac{k_1}{2(1-k_2)}$. The coefficient α_1 and β_1 are only related to the stress-strain curve of concrete, which forms the equivalent rectangle. The coefficient α_1 is the ratio between the constant stress value in the rectangular pattern and the design value of the compressive strength; The coefficient β_1 is the ratio between the height of the rectangle and the height of neutral axis x_c. The value of β_1 is: for $f_{cu,k} \leqslant 50 \text{N/mm}^2$, $\beta_1 = 0.8$; for $f_{cu,k} = 80 \text{N/mm}^2$, $\beta_1 = 0.74$; for $50 \text{N/mm}^2 < f_{cu,k} < 80 \text{N/mm}^2$, it is used in line interpolation, as shown in Table 3-1.

Values of ε_{cu}, β_1, α_1 and ξ_b Table 3-1

Concrete grade	ε_{cu}	β_1	α_1	ξ_b			
				HPB300	HRB335	HRB400	HRB500
≤C50	0.0033	0.80	1.0	0.576	0.550	0.518	0.482
C55	0.00325	0.79	0.99	0.566	0.541	0.508	0.473
C60	0.0032	0.78	0.98	0.556	0.531	0.499	0.464
C65	0.00315	0.77	0.97	0.549	0.522	0.490	0.455
C70	0.0031	0.76	0.96	0.537	0.512	0.481	0.447
C75	0.00305	0.75	0.95	0.526	0.503	0.472	0.438
C80	0.003	0.74	0.94	0.518	0.493	0.463	0.429

3.3.3 Critical Reinforcement Ratio

The boundary between the under reinforced beam and the over reinforced beam is called the "balanced beam". The balanced beam is the boundary that the steel bars in tension area yields and the edge fiber of the compression zone reaches its ultimate strain value at the same time, indicating the section is damaged. In Figure 3-11, take the strain of the steel bar as ε_y when it begins to yield.

$$\varepsilon_y = \frac{f_y}{E_s} \quad (3-9)$$

Where E_s——the elastic modulus of reinforcement.

The neutral axis height x_{cb} can be obtained from:

$$\frac{x_{cb}}{h_0} = \frac{\varepsilon_{cu}}{\varepsilon_{cu} + \varepsilon_y} \quad (3-10)$$

Take $x_b = \beta_1 x_{cb}$ into (3-10),

$$\frac{x_b}{\beta_1 h_0} = \frac{\varepsilon_{cu}}{\varepsilon_{cu} + \varepsilon_y} \quad (3-11)$$

When $\xi_b = \dfrac{x_b}{h_0}$, then

$$\xi_b = \frac{\beta_1}{1 + \dfrac{f_y}{E_s \varepsilon_{cu}}} \quad (3-12)$$

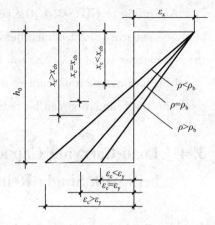

Figure 3-11 Height of boundary compression area

Where h_0——effective height of cross section;

x_b——balanced depth of compression zone;

f_y——value of tensile strength of longitudinal steel bar;

ε_{cu}——the ultimate strain value of concrete.

For $\xi > \xi_b$, it belongs to the over reinforced beam. The corresponding longitudinal tensile steel reinforcement ratio for balanced beam is called critical reinforcement ratio ρ_b, which can be calculated by Eqs. (3-13)~(3-14).

$$\alpha_1 f_c b x_b = f_y A_s \quad (3-13)$$

So

$$\rho_b = \frac{A_s}{bh_0} = \alpha_1 \xi_b \frac{f_c}{f_y} \quad (3-14)$$

The subscript b is the meaning of boundary.

3.3.4 Minimum Reinforcement Ratio

The character of the lightly reinforced beam is that once it cracks, the steel bars will rapidly yield to failure. To avoid this phenomenon, the minimum reinforcement ratio of the longitudinal tensile steel should be ensured, which is obtained by calculating the flexural bearing capacity of the plain concrete. To be noted, the whole height of the section h should be adopted instead of h_0. Considering the dispersion of the tensile strength and the influence of the shrinkage factors, the minimum reinforcement rate is usually based on previous experiences. In order to prevent the failure of lightly reinforced beams, the reinforcement ratio of the girder should be greater than $\rho_{min} h / h_0$.

According to GB 50010, the percentage of the reinforcement bars in the flexural members, eccentrically loaded members and the axial tension components should not be less than the larger value of 0.2% and 0.45 f_t/f_y.

In addition, the minimum reinforcement ratio of the tensile steel bars in the slabs directly on the foundation can be reduced properly, but not less than 0.15%.

3.4 Load-Carrying Capacity of Normal Section for Flexural Members with Singly Reinforced Rectangular Sections

3.4.1 Basic Formulae and Applicable Conditions

1. Basic formulae

Figure 3-12 shows the calculation diagram for a rectangular section only with tensile reinforcement, i.e., a singly reinforced rectangular section with an equivalent rectangle stress block at the ultimate state. In the figure, x is called the depth of concrete compression zone, z is called the internal lever arm.

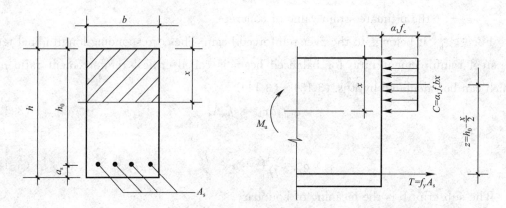

Figure 3-12 Rectangular section with tensile reinforcement

Considering the force equilibrium $\sum X = 0$
$$f_y A_s = \alpha_1 f_c b x \tag{3-15}$$
Considering the bending moment equilibrium $\sum M = 0$
$$M \leqslant M_u = \alpha_1 f_c b x \left(h_0 - \frac{x}{2}\right) = f_y A_s \left(h_0 - \frac{x}{2}\right) \tag{3-16}$$

2. Applicable conditions

In order to prevent the occurrence of over-reinforced failure:
$$\rho \leqslant \rho_{max} = \xi_b \frac{\alpha_1 f_c}{f_y} \tag{3-17}$$

or
$$x \leqslant \xi_b h_0 \tag{3-18}$$

Obviously, by substituting $\xi = \xi_b$ into Eq. (3-18), the maximum flexural bearing capacity of under-reinforced sections can be expressed as

$$M_{u,\max} = \alpha_1 f_c b h_0^2 \xi_b (1 - 0.5\xi_b) \tag{3-19}$$

In order to prevent the occurrence of lightly-reinforced failure

$$A_s \geqslant \rho_{\min} bh \tag{3-20}$$

or

$$\rho \geqslant \rho_{\min} \frac{h}{h_0} \tag{3-21}$$

3.4.2 Application of Formulae

1. Cross-sectional design

For this kind of problem, the external bending moment M of the normal section is designed to be equal to the design value of the bending bearing capacity M_u, that is, $M = M_u$.

Usually the cross-sectional dimensions (b, h and h_0), material properties (f_c, f_t and f_y) and the external bending moment M are already known to calculate the reinforcement content (A_s).

The calculation should be carried out as follows:

(1) Calculate the effective depth h_0 of the cross section: $h_0 = h - a_s$, the effective depth of the cross section can be estimated according to the following principles for members in the first type of environment:

For beams of single-row steel bars, $h_0 = h - 40$mm;

For beams of double-row steel bars, $h_0 = h - 65$mm;

For slabs, $h_0 = h - 20$mm.

(2) Calculate x by Eq. (3-16).

(3) If $x \leqslant \xi_b h_0$, A_s can be obtained from Eq. (3-15). If $x > \xi_b h_0$, the following measures can be taken as to increase the cross-sectional dimensions or the grade of concrete strength, or increase to doubly reinforced sections.

(4) The A_s obtained from step (3) should also be greater than $\rho_{\min} bh$. If not, take $A_s = \rho_{\min} bh$ as the final value.

2. Evaluation of bearing capacity of cross-section

Knowing the cross-sectional dimensions (b and h), reinforcement (A_s), and material properties (f_c, f_t and f_y), the flexural bearing capacity M_u can be evaluated according to the following steps:

(1) Calculate the effective depth h_0 of the cross section: $h_0 = h - a_s$, which can be esti-

mated according to the following principles:

For beams of single-row steel bars, $h_0 = h - c - d_s - d/2$;

For beams of double-row steel bars, $h_0 = h - c - d_s - d - \max(d/2, 25/2)$;

Where c is the thickness of the concrete cover, d_s and d are the diameters of the stirrups and steel bars respectively.

(2) Calculate x from Eq. (3-15).

(3) If $x \leqslant \xi_b h_0$, calculate M_u from Eq. (3-16).

(4) If $x > \xi_b h_0$, $M_u = M_{u,\max} = \alpha_1 f_c b h_0^2 \xi_b (1 - 0.5\xi_b)$.

3.4.3 Calculation Coefficient and Calculation Method on Bearing Capacity of Normal Section

Take calculation coefficient as

$$\alpha_s = \frac{M}{\alpha_1 f_c b h_0^2} \tag{3-22}$$

That is $\alpha_s = \xi(1 - 0.5\xi)$.

Take calculation coefficient as

$$\gamma_s = \frac{z}{h_0} \tag{3-23}$$

That is $\gamma_s = 1 - 0.5\xi$. Take $M = M_u$, the following equations can be obtained by solving Eqs. (3-16) and (3-17).

$$\xi = 1 - \sqrt{1 - 2\alpha_s} \tag{3-24}$$

$$\gamma_s = \frac{1 + \sqrt{1 - 2\alpha_s}}{2} \tag{3-25}$$

Therefore, α_s is obtained by the formula (3-22), subsequently the values of ξ and γ_s can be obtained from Eqs. (3-23) and (3-24). γ_s reflects the ratio of the internal lever arm to the effective depth of the section which is called the internal lever arm coefficient of the internal torque, and α_s is called the section resistance coefficient.

In cross-sectional design, the internal lever arm coefficient is obtained, thus the cross-sectional area of the longitudinal tensile reinforcement can be easily calculated as:

$$A_s = \frac{M}{\gamma_s f_y h_0} \tag{3-26}$$

[**Example 3-1**] The design bending moment for a reinforcement concrete beam of width $b = 250$mm and depth $h = 550$mm is $M = 200$kN·m. The concrete is taken as grade C25, and the steel is of grade HRB335. The calculation for the design of tensile reinforcements is required.

[**Data**] C25 concrete: $f_c = 11.9$N/mm^2, $\alpha_1 = 1.0$, $f_t = 1.27$N/mm^2

HRB335 steel: $f_y = 300\text{MPa}$, $\xi_b = 0.55$

Effective depth: $h_0 = 550 - a_s = 550 - 35 = 515\text{mm}$

Minimum steel ratio $\rho_{min} = 0.2\% > 0.45 f_t/f_y = 0.19\%$

[Solution]

$$\alpha_s = \frac{M}{\alpha_1 f_c b h_0^2} = \frac{200 \times 10^6}{1 \times 11.9 \times 250 \times 515^2} = 0.25$$

$$\xi = 1 - \sqrt{1 - 2\alpha_s} = 1 - \sqrt{1 - 2 \times 0.25} = 0.293 < \xi_b = 0.55$$

$$A_s = \frac{\alpha_1 f_c b h_0 \xi}{f_y} = \frac{1 \times 11.9 \times 250 \times 515 \times 0.293}{300} = 1496.4\text{mm}^2$$

and $A_s > \rho_{min} b h = 0.002 \times 250 \times 550 = 275\text{mm}^2$

The result is to adopt 4 Φ 22 with a total steel area of 1520mm².

[**Example 3-2**] The design bending moment for a rectangular section beam with span $l = 6$m is $M = 160\text{kN} \cdot \text{m}$. The concrete is of grade C25 and the steel is of grade HRB400. The selection of section sizes b and h, and the design for tensile reinforcements are required.

[Data] C25 concrete: $f_c = 11.9\text{N/mm}^2$, $\alpha_1 = 1.0$, $f_t = 1.27\text{N/mm}^2$

HRB400 steel: $f_y = 360\text{MPa}$, $\xi_b = 0.518$

Minimum steel ratio: $\rho_{min} = 0.2\% > 0.45 f_t/f_y = 0.16\%$

[Solution]

(1) Select the dimensions of the section, b and h:

$$h = \left(\frac{1}{14} \sim \frac{1}{10}\right) l = 430 \sim 600\text{mm}$$

Take $h = 500\text{mm}$, $b = 200\text{mm}$, $h_0 = 500 - 35 = 465\text{mm}$.

(2) Calculate the area A_s of the reinforcements

$$\alpha_s = \frac{M}{\alpha_1 f_c b h_0^2} = \frac{160 \times 10^6}{1 \times 11.9 \times 200 \times 465^2} = 0.311$$

$$\xi = 1 - \sqrt{1 - 2\alpha_s} = 1 - \sqrt{1 - 2 \times 0.311} = 0.385 < \xi_b = 0.518$$

$$A_s = \frac{\alpha_1 f_c b h_0 \xi}{f_y} = \frac{1 \times 11.9 \times 200 \times 465 \times 0.385}{360} = 1183.6\text{mm}^2$$

and $A_s > \rho_{min} b h = 0.002 \times 200 \times 500 = 200\text{mm}^2$

The result is to adopt 4 Φ 20 bars with a total steel area of 1256mm².

3.5 Load-Carrying Capacity of Normal Section for Flexural Members with Doubly Reinforced Rectangular Sections

3.5.1 Introduction

In practical concrete engineering, the number of longitudinal compression reinforcements arranged in the compression zone should be enough, acting as the role of steel frame and the role of components for bearing capacity. This section type is called the doubly reinforced section. Because it is not economical to bear compression forces by steels, the doubly reinforced section is only available in the following cases:

(1) The section is subjected to a reversal bending moment under different load combinations;

(2) The section is unable to suffer more bending moments due to its limited depth for architectural reasons.

3.5.2 Basic Formulae and Applicable Conditions

1. The value of compressive strength of longitudinal compression steels

According to the strain distribution of an under-reinforced section (Figure 3-13), the strain of compression reinforcement can be obtained as follows:

Take

$$x = 2a'_s, \ \varepsilon_{cu} \approx 0.0033, \ \beta_1 \approx 0.8. \tag{3-27}$$

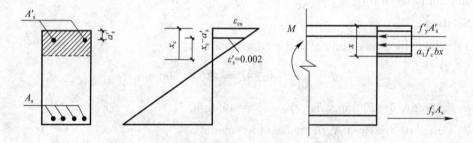

Figure 3-13 The strain and stress of the compression steels

The strain can be calculated as

$$\varepsilon'_s = 0.0033\left(1 - \frac{0.8a'_s}{x}\right) = 0.0033\left(1 - \frac{0.8a'_s}{2a'_s}\right)$$

$$= 0.00198 \approx 0.002 \tag{3-28}$$

Thus

$$\sigma'_s = E_s \varepsilon'_s = (1.95 \sim 2.1) \times 10^5 \times 0.002$$

$$= 390 \sim 420 \text{N/mm}^2 \tag{3-29}$$

Where a'_s——the distance from the centroid of the compression reinforcement to the compressive edge of the section.

For the steel bars of 300MPa, 335MPa and 400MPa, the value of σ'_s has exceeded the yield strength f'_y. Thus compression steels can achieve their yield strength f'_y when $x \geqslant 2a'_s$.

In addition, the stirrups in flexural members of doubly reinforced sections should satisfy the following details: (1) The stirrups should be closed with the spacing no more than 15d and 400mm; (2) The diameter of the stirrup should be at least $d/4$, where d is the maximum diameter of the longitudinal reinforcement.

2. Basic formulae and applicable conditions

Figure 3-14 shows the calculation diagram for a rectangular section with doubly reinforced section.

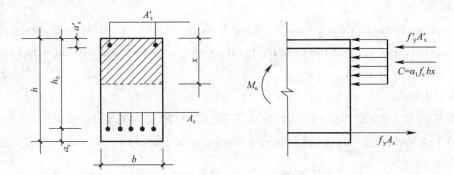

Figure 3-14　Rectangular section with compression reinforcement

Considering the force equilibrium $\Sigma X = 0$
$$f_y A_s = \alpha_1 f_c bx + f'_y A'_s \tag{3-30}$$
Considering the bending moment equilibrium $\Sigma M = 0$
$$M \leqslant M_u = \alpha_1 f_c bx \left(h_0 - \frac{x}{2}\right) + f'_y A'_s (h_0 - a'_s) \tag{3-31}$$
When applying the above two formulae, the following applicable conditions must be met:
(1) $x \leqslant \xi_b h_0$
(2) $x \geqslant 2a'_s$

When the condition (2) is not satisfied, taking moments to the point of A'_s, normal sectional bending capacity is calculated as follows:
$$M_u = f_y A_s (h_0 - a'_s) \tag{3-32}$$

3.5.3　Application of Formulae

1. Normal sectional design

The normal sectional design based on the flexural capacity can be classified into two cases:

Case 1:

The cross-sectional dimensions (b, h and h_0), material properties (f_c, f_t and f_y), and the external bending moment M are already known to calculate the reinforcement areas

(A_s, A_s').

The calculation can be done as follows:

(1) In order to fully utilize the concrete in the compression zone, $x=\xi_b h_0$ is taken;

(2) Calculate A_s' from Eq. (3-31);

(3) Calculate A_s from Eq. (3-30).

Since $x=\xi_b h_0$ has been assumed in the first step, all the applicable conditions are automatically satisfied.

Case 2:

The cross-sectional dimensions (b, h and h_0), material properties (f_c, f_t and f_y), the compression steels (A_s') and the external bending moment M are already known to calculate the reinforcement area (A_s).

(1) Calculate x from Eq. (3-31);

(2) If $2a_s' \leqslant x \leqslant \xi_b h_0$, calculate A_s from Eq. (3-30). If $x > \xi_b h_0$, the area for compression steels (A_s') should be increased and following steps are calculated by case 1. If $x < 2a_s'$, calculate A_s from Eq. (3-32).

2. Evaluation of bearing capacity of cross-section

The cross-sectional dimensions (b, h and h_0), reinforcement areas (A_s, A_s') and material properties (f_c, f_t and f_y) are known, the flexural bearing capacity M_u can be evaluated according to the following steps:

(1) Calculate x from Eq. (3-30);

(2) If $2a_s' \leqslant x \leqslant \xi_b h_0$, calculate M_u from Eq. (3-31);

(3) If $x > \xi_b h_0$, $M_u = M_{u,\max} = \alpha_1 f_c b h_0^2 \xi_b (1-0.5\xi_b) + f_y' A_s' (h_0 - a_s')$;

(4) If $x \leqslant 2a_s'$, calculate M_u from Eq. (3-32).

[Example 3-3] A rectangular section with width $b=250$mm and depth $h=600$mm is to carry a design bending moment $M=310$kN·m. The concrete is of grade C30. The section is reinforced with 3⏀18 as compression steels of HRB400. The tension reinforcement A_s is required.

[Data] C30 concrete: $f_c=14.3$N/mm^2, $\alpha_1=1.0$, $f_t=1.43$N/mm^2

HRB400 steel: $f_y'=f_y=360$MPa, $\xi_b=0.518$

Steel 3⏀18: $A_s'=763$mm^2, $a_s'=35$mm

Effective depth: $h_0=h-a_s=600-60=540$mm

[Solution]

$$\alpha_s = \frac{M - f_y' A_s'(h_0 - a_s')}{\alpha_1 f_c b h_0^2}$$

$$= \frac{310 \times 10^6 - 360 \times 763 \times (540-35)}{1 \times 14.3 \times 250 \times 540^2} = 0.164$$

$$\xi = 1 - \sqrt{1-2\alpha_s} = 0.18 < \xi_b = 0.518, \quad x = \xi h_0 = 97.2\text{mm} > 2a'_s = 70\text{mm}$$

$$A_s = \frac{\alpha_1 f_c b h_0 \xi + f'_y A'_s}{f_y} = 1728.3\text{mm}^2$$

The result is to adopt $4 \phi 20 + 2 \phi 18 (A_s = 1765\text{mm}^2)$ bars for tension reinforcement.

[**Example 3-4**] The sectional size of beam is $b \times h = 200\text{mm} \times 500\text{mm}$; the concrete is of grade C40; the section is reinforced with steel of HRB400; and the bending moment is $M = 330\text{kN} \cdot \text{m}$. The environment category is Class 1.

The sectional areas A_s and A'_s of the tension reinforcement and compression reinforcement are required.

[**Solution**]

$f_c = 19.1\text{N/mm}^2$, $f_y = f'_y = 360\text{N/mm}^2$

$\alpha_1 = 1.0$, $\beta_1 = 0.8$

Assuming that the tensioned bars are arranged in two layers, set $a_s = 65\text{mm}$, then

$$h_0 = h - a_s = 500 - 65 = 435\text{mm}$$

$$\alpha_s = \frac{M}{\alpha_1 f_c b h_0^2} = \frac{330 \times 10^6}{1 \times 19.1 \times 200 \times 435^2} = 0.457$$

$$\xi = 1 - \sqrt{1-2\alpha_s} = 0.707 > \xi_b = 0.518$$

This shows that if it is designed into a rectangular section with single-row reinforcing bars, the over-reinforcement situation will occur with $x > \xi_b h_0$. If there are restraints to increase the cross-sectional size or improve the strength grade of concrete, it should be designed into a rectangular section with double-row reinforcing bars.

Take $\xi = \xi_b$, obtained from Eq. (3-16):

$$M_{u2} = \alpha_1 f_c b h_0^2 \xi_b (1 - 0.5\xi_b)$$
$$= 1.0 \times 19.1 \times 200 \times 435^2 \times 0.518 \times (1 - 0.5 \times 0.518)$$
$$= 277.45\text{kN} \cdot \text{m}$$

$$A'_s = \frac{M - M_{u2}}{f'_y (h_0 - a'_s)} = \frac{330 \times 10^6 - 277.45 \times 10^6}{360 \times (435 - 40)} = 370\text{mm}^2$$

Obtained from Eq. (3-30)

$$A_s = \xi_b \frac{\alpha_1 f_c b h_0}{f_y} + A'_s$$

$$= 0.518 \times \frac{1 \times 19.1 \times 200 \times 435}{360} + 370$$

$$= 2761\text{mm}^2$$

The result is to choose $3 \phi 25 + 1 \phi 25 + 2 \phi 22$ bars for tension reinforcement, of which the total area A_s is 2724mm^2, $2 \phi 16$ bars are chosen for compression reinforcement, A'_s of which is 402mm².

[**Example 3-5**] Given a rectangular section beam of width $b=250$mm and depth $h=550$mm. The concrete is of grade C25. The section is reinforced with compression steel 2 ⏀ 16 and tension steel 3 ⏀ 22 of HRB335. The ultimate moment M_u of the section is required.

[**Data**] C25 concrete: $f_c=11.9\text{N/mm}^2$, $\alpha_1=1.0$

HRB335 steel: $f_y=300\text{MPa}$, $\xi_b=0.518$

Steel area 2 ⏀ 16: $A'_s=402\text{mm}^2$; 3 ⏀ 22: $A_s=1140\text{mm}^2$

Effective depth $h_0=h-(c+d/2)=550-(25+22/2)=514$mm

[**Solution**]

$$\xi = \frac{f_y A_s - f'_y A'_s}{\alpha_1 f_c b h_0} = \frac{300 \times 1140 - 300 \times 402}{1 \times 11.9 \times 250 \times 514} = 0.145 < \xi_b$$

$$x = \xi h_0 = 74.5 > 2a'_s = 70\text{mm}$$

$$M_u = \alpha_1 f_c b x \left(h_0 - \frac{x}{2}\right) + f'_y A'_s (h_0 - a'_s)$$

$$= 1 \times 11.9 \times 250 \times 74.5 \times \left(514 - \frac{74.5}{2}\right) + 300 \times 402 \times (514-35) \times 10^{-6}$$

$$= 163.4\text{kN} \cdot \text{m}$$

3.6 Load-Carrying Capacity of Normal Section for Flexural Members with T-Sections

3.6.1 Introduction

When the flexural member is broken, most of the concrete in tensile zone have already been out of work, therefore the redundant concrete in the tension zone can be excavated (shown in the shaded part in Figure 3-15a), but the calculation on flexural bearing capacity of normal section is identical with the original rectangular section bearing capacity, leading to the saving on concrete and reduction on self-weight. The rest of the beam is called T-shaped section consisting of beam rib and cantilever flange as two parts.

However, if the flanges locate in the tension zone of the beam, as shown in Figure 3-15(b), the concrete in tensile zones of the inverted T-shaped section beam will all quit work at the final stage. Thus, the bending bearing capacity should be calculated as the rectangular cross section with the width of b.

A typical T-section is composed of two parts, namely, the flange and web as shown in Figure 3-16. The width of the flange is b'_f, the thickness of the flange is h'_f, the width of the web is b, and the depth of the section is h.

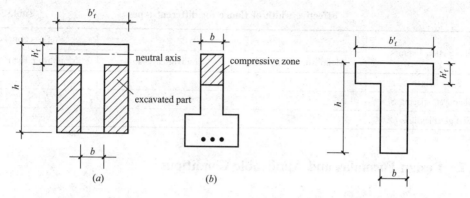

Figure 3-15 The members of T-section　　　　**Figure 3-16 T-section**

According to the experimental and theoretical analysis, the longitudinal compressive stress on the flange is not evenly distributed along the width of the flange when subjected to loads, and the stress farther from the web shows smaller values. According to the elastic mechanics, the distribution of the compressive stress depends on the relative size and the loading form of the cross section. However, due to the development of plastic deformation, the actual compressive stress distribution is more uniform than that of elasticity analysis (shown in Figure 3-17a, c). For some cast-in-place T-section beams with wide flange, the compressive stress far from the web is too small. Therefore, a substitutive method can be conducted to take a certain range of width into calculation as the effective width b'_f, where the compressive stress is uniformly distributed as shown in Figure 3-17 (b) and (d).

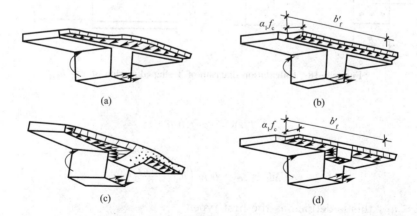

Figure 3-17 The actual stress and calculation of the compression zone of the T-section beams
(a), (c) actual stress; (b), (d) calculation stress

The effective width b'_f of a flange, specified by GB 50010, is identified for each type, as listed in Table 3-2.

Effective width of flange for different types Table 3-2

Restriction Factors	T-section		Γ-section
	Monolithic beam-slab	Individual beam	Monolithic beam-slab
Span length l_0	$l_0/3$	$l_0/3$	$l_0/6$
Clear web spacing S_n	$b+S_n$	—	$b+S_n/2$
Flange thickness h'_f	$b+12h'_f$	b	$b+5h'_f$

3.6.2 Design Formulas and Applicable Conditions

When calculating the T-shaped beam, it can be divided into two types according to the neutral axis position.

Figure 3-18 shows the stress distribution of a T-shaped section of $x=h'_f$. The equilibrium equation of force is available as:

$$f_y A_s = \alpha_1 f_c b'_f h'_f \tag{3-33}$$

The equilibrium equation of the bending moment is available as:

$$M_u = \alpha_1 f_c b'_f h'_f \left(h_0 - \frac{h'_f}{2}\right) \tag{3-34}$$

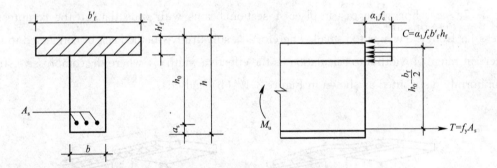

Figure 3-18 Calculation diagram of T-shaped section of $x=h'_f$

If

$$f_y A_s \leqslant \alpha_1 f_c b'_f h'_f \tag{3-35}$$

Or

$$M_u \leqslant \alpha_1 f_c b'_f h'_f \left(h_0 - \frac{h'_f}{2}\right) \tag{3-36}$$

$x \leqslant h'_f$, this is defined as the first type.

If

$$f_y A_s > \alpha_1 f_c b'_f h'_f \tag{3-37}$$

Or

$$M_u > \alpha_1 f_c b'_f h'_f \left(h_0 - \frac{h'_f}{2}\right) \tag{3-38}$$

$x > h'_f$, this is defined as the second type.

1. T-section of the first type ($x \leqslant h'_f$)

Figure 3-19 shows the stress distribution of a T-section of the first type. The actual stress distribution is replaced into an equivalent stress block in a similar way as conducted in the rectangular section analysis, the equilibrium equation is established as

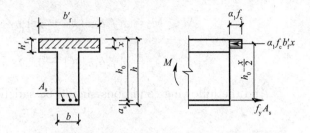

Figure 3-19 Stress calculation for the first type of T-section

$$f_y A_s = \alpha_1 f_c b'_f x \qquad (3\text{-}39)$$

$$M \leqslant M_u = \alpha_1 f_c b'_f x \left(h_0 - \frac{x}{2}\right) = \alpha_1 f_c b'_f h_0^2 \xi (1 - 0.5)\xi \qquad (3\text{-}40)$$

And the following conditions should be satisfied:

$$A_s \geqslant \rho_{\min} bh \qquad (3\text{-}41)$$

Generally the applicable condition is automatically satisfied as $\xi \leqslant \xi_b$.

2. T-section of the section type ($x > h'_f$)

Figure 3-20 shows the stress distribution of the T-section of the second type. The equilibrium equations are established as:

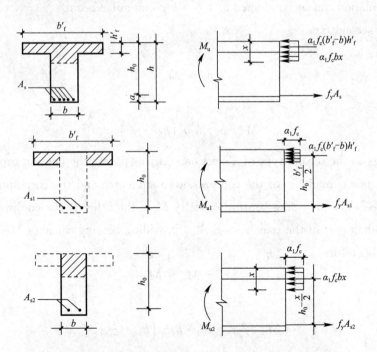

Figure 3-20 Stress calculation for the second type of T-section

$$f_y A_s = \alpha_1 f_c bx + \alpha_1 f_c (b'_f - b) h'_f \qquad (3\text{-}42)$$

$$M \leqslant M_u = \alpha_1 f_c b_f' x \left(h_0 - \frac{x}{2}\right) + \alpha_1 f_c (b_f' - b)\left(h_0 - \frac{h_f'}{2}\right)$$

$$= \alpha_1 f_c b_f' h_0^2 \xi(1-0.5)\xi + \alpha_1 f_c (b_f' - b)\left(h_0 - \frac{h_f'}{2}\right) \quad (3\text{-}43)$$

And the following conditions should be satisfied:

$$x \leqslant \xi_b h_0 \quad (3\text{-}44)$$

Generally the applicable condition are automatically satisfied as $\rho \geqslant \frac{h}{h_0}\rho_{\min}$.

3.6.3 Application of Formulae

1. Normal sectional design

The general section size is known and the tensile reinforcement area A_s is required, which can be calculated as follows under two different types:

(1) The first type

Take

$$M = M_u$$

If

$$M \leqslant \alpha_1 f_c b_f' h_f' \left(h_0 - \frac{h_f'}{2}\right)$$

The calculation can be conducted as a $b_f' \times h$ rectangular beam.

(2) The second type

Take

$$M = M_u$$

If

$$M > \alpha_1 f_c b_f' h_f' \left(h_0 - \frac{h_f'}{2}\right)$$

It belongs to the second type of T-section. At this point, the bearing capacity consists of two parts: one is made up of the compression flange area and the corresponding part of the tensile steel A_{s1}, providing bearing capacity M_{u1}; the other is the compression rib and the corresponding part of the tensile steel A_{s2}, providing bearing capacity M_{u2}, as shown in Figure 3-20. Therefore

$$M_u = M_{u1} + M_{u2} \quad (3\text{-}45)$$

And

$$M_{u1} = \alpha_1 f_c (b_f' - b) h_f' \left(h_0 - \frac{h_f'}{2}\right) \quad (3\text{-}46)$$

$$M_{u2} = \alpha_1 f_c b x \left(h_0 - \frac{x}{2}\right) \quad (3\text{-}47)$$

As shown in Figure 3-20

$$A_{s1} = \frac{\alpha_1 f_c (b'_f - b) h'_f}{f_y} \tag{3-48}$$

Summing up A_{s2} calculated by the rectangular section with singly reinforced bars:

$$A_s = A_{s1} + A_{s2} = \frac{\alpha_1 f_c (b'_f - b) h'_f}{f_y} + A_{s2} \tag{3-49}$$

Finally, the compressive height x should be examined to meet $x \leqslant \xi_b h_0$.

2. Evaluation for bearing capacity of normal section

If $f_y A_s \leqslant \alpha_1 f_c b'_f h'_f$, the section is analyzed as the first type T-section; M_u can be calculated as a $b'_f \times h$ rectangular beam.

If $f_y A_s > \alpha_1 f_c b'_f h'_f$, the section is to be analyzed as the second type T-section, the bearing capacity can be calculated as follows:

(1) Calculate A_{s1}

$$A_{s1} = \frac{\alpha_1 f_c (b'_f - b) h'_f}{f_y} \tag{3-50}$$

(2) Calculate A_{s2}

$$A_{s2} = A_s - A_{s1} \tag{3-51}$$

(3) Calculate x

$$x = \frac{f_y A_{s2}}{\alpha_1 f_c b} \tag{3-52}$$

(4) Calculate M_{u1} and M_{u2}

$$M_{u1} = \alpha_1 f_c (b'_f - b) h'_f \left(h_0 - \frac{h'_f}{2} \right) \tag{3-53}$$

$$M_{u2} = \alpha_1 f_c b x \left(h_0 - \frac{x}{2} \right) \tag{3-54}$$

(5) The summation of the bearing capacity

$$M_u = M_{u1} + M_{u2} \tag{3-55}$$

(6) The external bending moment should satisfy:

$$M_u \geqslant M \tag{3-56}$$

[**Example 3-6**] The design bending moment for a T-section as shown in Figure 3-21 is $M = 350$ kN · m. If the concrete is of grade C30 and the steel bar is of grade HRB400, the design for tensile reinforcement is required.

[**Data**] C30 concrete: $f_c = 14.3 \text{N/mm}^2$, $\alpha_1 = 1.0$, $f_t = 1.43 \text{N/mm}^2$

HRB400 steel: $f_y = 360 \text{MPa}$, $\xi_b = 0.518$

Effective depth: $h_0 = 600 - a_s = 600 - 60 = 540 \text{mm}$

[**Solution**]

$$M = 350 \text{kN} \cdot \text{m} > \alpha_1 f_c b'_f h'_f \left(h_0 - \frac{h'_f}{2} \right) = 286 \text{kN} \cdot \text{m}$$

The T-section belongs to the second type, thus

$$\alpha_s = \frac{M - \alpha_1 f_c (b'_f - b) h'_f \left(h_0 - \frac{h'_f}{2}\right)}{\alpha_1 f_c b h_0^2}$$

$$= \frac{350 \times 10^6 - 14.3 \times (500 - 250) \times 80 \times \left(540 - \frac{80}{2}\right)}{14.3 \times 250 \times 540^2} = 0.199$$

$$\xi = 1 - \sqrt{1 - 2\alpha_s} = 0.224 < \xi_b = 0.518$$

$$A_s = \frac{\alpha_1 f_c b h_0 \xi + \alpha_1 f_c (b'_f - b) h'_f}{f_y}$$

$$= \frac{14.3 \times 250 \times 540 \times 0.224 + 14.3 \times (500 - 250) \times 80}{360} = 1995.6 \text{mm}^2$$

The design is to adopt 4⌽22+2⌽18 bars with the area of $A_s = 2029 \text{mm}^2$ and the arrangement is shown in Figure 3-21.

[**Example 3-7**] Given a T-section as shown in Figure 3-22, the concrete is of C20 and the reinforcements are 6⌽22 bars of grade HRB335. The ultimate bending moment M_u is required.

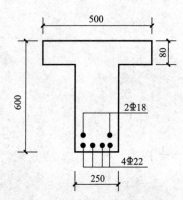

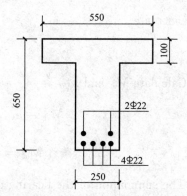

Figure 3-21 Beam section for Example 3-6   Figure 3-22 Beam section for Example 3-7

[**Data**] C20 concrete: $f_c = 9.6 \text{N/mm}^2$, $\alpha_1 = 1.0$
HRB335 steel: $f'_y = f_y = 300 \text{MPa}$, $\xi_b = 0.55$
Effective depth: $h_0 = h - a_s = 650 - 60 = 590 \text{mm}$

[**Solution**]

$$f_y \times A_s = 300 \times 2281 = 684300 \text{N} > \alpha_1 f_c b'_f h'_f = 9.6 \times 550 \times 100 = 528000 \text{N}$$

The T-section belongs to the second type. Thus

$$x = \frac{f_y A_s - \alpha_1 f_c (b'_f - b) h'_f}{\alpha_1 f_c b} = \frac{684300 - 9.6 \times (550 - 250) \times 100}{9.6 \times 250}$$

$$= 165.1 \text{mm} < \xi_b h_0 = 324.5 \text{mm}$$

$$M \leqslant M_u = \alpha_1 f_c b x \left(h_0 - \frac{x}{2}\right) + \alpha_1 f_c (b'_f - b) \left(h_0 - \frac{h'_f}{2}\right)$$

$$= 9.6 \times 250 \times 165.125 \times \left(590 \times \frac{165.125}{2}\right) \times 10^{-6} + 9.6 \times (550 - 250)$$
$$\times \left(590 \times \frac{100}{2}\right) \times 10^{-6} = 202.5 \text{kN} \cdot \text{m}$$

Questions

3-1 What state does the edge concrete tension fiber reach when the tension zone is assumed to be cracked?

3-2 What is a balanced failure? What are the meanings of "ε_s" and "ε_{cu}" under a balanced failure?

3-3 What are the flexural failure modes for reinforced concrete beams and how are they classified?

3-4 What are the reinforcement details in reinforced concrete slabs?

3-5 How is the damage of an under reinforced concrete or over reinforced concrete beam determined by flexure?

3-6 What principle is used to determine the minimum reinforcement ratio of the longitudinal steel bars in reinforced concrete beams?

3-7 If the longitudinal steel bars are uniformly distributed across a reinforced concrete rectangular section, is the flexural bearing capacity calculated by $M_u = A_s f_y (h_0 - x/2)$ and is that calculation consistent with the actual flexural bearing capacity? Why?

3-8 During the design of a doubly reinforced rectangular section, how does one ensure that the section fails at the same time as the longitudinal steels yield?

3-9 How is the flexural bearing capacity of a doubly reinforced rectangular section calculated if $x < 2A_s'$?

3-10 Why should the effective width of the compression flange of a T-section be specified?

3-11 How do the flexural bearing capacity and the ductility of reinforced concrete beams change with the increase of the longitudinal bars?

Exercises

3-1 A rectangular section with width $b = 250$mm and depth $h = 500$mm is to carry a design bending moment $M = 260$kN $\cdot$ m. The concrete is of grade C30 and the tensile steel is of HRB400. The tension reinforcement A_s is required.

3-2 A rectangular section with width $b=200$mm and depth $h=450$mm is to carry a design bending moment $M=145$kN · m. The concrete is of grade C40 and the tensile steel is of HRB400. The tension reinforcement A_s is required.

3-3 A rectangular section with width $b=200$mm and depth $h=450$mm is reinforced with 4 ⌽ 16 of HRB400. The design bending moment is $M=100$kN · m. Is the section safe or not?

3-4 A doubly reinforced section of $b=250$mm and $h=500$mm is to carry a design bending moment $M=260$kN · m. The concrete is of grade C25 and the steel bars are of HRB335. The tension reinforcement A_s and the compression reinforcement A'_s are required.

3-5 A T-section beam with $b'_f=550$mm, $b=250$mm, $h=750$mm and $h'_f=100$mm is shown in Figure 3-23. The concrete is of grade C40, the steel is of grade HRB335, the design moment is $M=500$kN · m. The tensile reinforcement A_s is required.

3-6 A section beam is with $b'_f=400$mm, $b=200$mm, $h=500$mm and $h'_f=80$mm. The concrete is of grade C30, the design moment is $M=270$kN · m and the steel is HRB400. The tension reinforcement A_s is required.

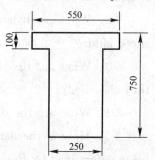

Figure 3-23 Exercise 3-5

3-7 A T-section beam is with $b'_f=600$mm, $b=300$mm, $h=700$mm and $h'_f=120$mm. The concrete is of grade C30, the design moment is $M=600$kN · m. The section is reinforced with tension steels of 8 ⌽ 22. Is the section safe or not?

Chapter 4
Load-Carrying Capacity of Oblique Section for Flexural Members

4.1 Introduction

It is risky that the flexural members of reinforced concrete are damaged along the inclined cracks in the section near the support by the shear force and bending moment. Therefore, the load-carrying capacity of normal section should be ensured, as well as the load-carrying capacity of the oblique section, which includes two aspects: the shear load-carrying capacity and the flexural load-carrying capacity. Usually for plates, the ratio of the span to the height is relatively large, and most of them are subjected to distributed load. Therefore, comparing with the load-carrying capacity of normal section, the capacity in the oblique section is often sufficient, so the load-carrying capacity of the oblique section of the flexural member is mainly focused on the beam and the thick plate.

For this purpose, transverse reinforcements, or called web reinforcements, should be used to provide the adequate ultimate strength for oblique sections. The web reinforcements are composed of transverse ties, stirrups and inclined bars bent up from the surplus of longitudinal reinforcing bars. The longitudinal bars bound together with the inclined bent-up bars form a fairly rigid skeleton of steel bars as shown in Figure 4-1.

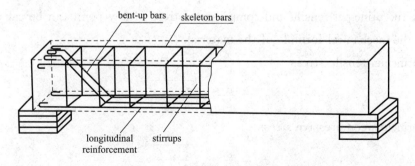

Figure 4-1 The stirrups and bent-up bars

Generally, stirrups should be bent into bent-up bars, but oblique stirrups are difficult to tie. Therefore, vertical reinforcement is usually adopted. The experimental research shows that the effect of stirrups on restraining oblique cracks is better than the bent-up bars. So in engineering design, the stirrups are used preferentially. Because the tension of the bent-up bars is relatively large and concentrated, it may cause splitting cracks in the bent concrete, as shown in Figure 4-2. Therefore, large amount use of bent-up bars should be avoided, and the angle should take 45° or 60°.

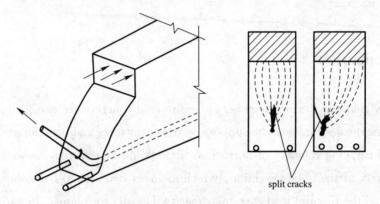

Figure 4-2 The split cracks

4.2 Inclined Crack, Shear-Span Ratio and Failure Modes of Inclined Sections under Shear Force

4.2.1 Web-Shear Inclined Crack and Flexure-Shear Inclined Crack

The reinforced concrete beams combined with the shear force and bending moment will produce oblique cracks in the flexure-shear section. The inclined cracks are mainly divided into two types: flexure-shear inclined crack and web-shear inclined crack. Before the beam is cracked, the principal tensile and compressive stress of any point can be calculated according to the mechanical formula of the material.

The principal tensile stress

$$\sigma_{tp} = \frac{\sigma}{2} + \sqrt{\frac{\sigma^2}{4} + \tau^2} \qquad (4-1)$$

The principal compressive stress

$$\sigma_{cp} = \frac{\sigma}{2} - \sqrt{\frac{\sigma^2}{4} + \tau^2} \qquad (4-2)$$

α is the angle between the direction of the principal tensile stress and the axis of the beam.

$$\tan(2\alpha) = -\frac{2\tau}{\sigma} \tag{4-3}$$

Figare 4-3 shows the trace of the principal stress of a beam without web reinforcement. The solid line is the principal tensile stress trace, and the dotted line is the principal compressive stress trace.

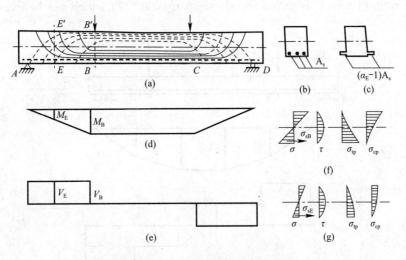

Figure 4-3 The trace of the principal stress

In the vicinity of the neutral axis, the principal tensile stress and the principal compressive stress are orthogonal and inclined at an angle of 45° from the longitudinal axis of a beam. When the load increases and the tensile strain reaches the ultimate tensile strain of concrete, the concrete cracks along the main compressive stress trace, which is called the web-shear inclined crack. The web-shear inclined crack is elliptical shape-wide in the middle and fine in both ends, which often appears in the T-shaped section with thin web as shown in Figure 4-4 (a). It can be seen from the principal stress traces that the principal tensile stress is horizontal at the lower edge of the section in the flexure-shear sections, and some shorter vertical cracks may occur and develop into inclined cracks extending to the concentrated load point, which are called the flexural-shear cracks. The flexural-shear crack is wide at the lower and fine on the top as shown in Figure 4-4 (b).

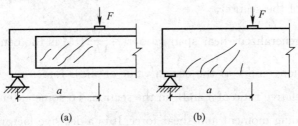

Figure 4-4 The inclined crack

(a) The web-shear inclined crack; (b) The flexure-shear inclined crack

4.2.2 Shear-Span Ratio (λ)

Figure 4-5 shows a simply supported beam under concentrated loads, a is the distance from the outermost concentrated force to the adjacent support, and h_0 is the effective depth. The ratio of a to h_0 is called the shear-span ratio (λ), which can be expressed as:

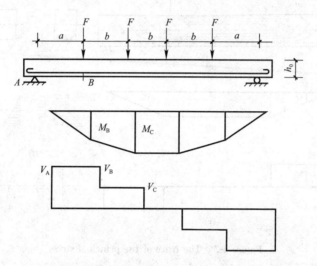

Figurt 4-5 Simply supported beam under concentrated loads

$$\lambda = \frac{a}{h_0} \tag{4-4}$$

For the rectangular section beam, the normal stress (σ) and the shear stress (τ) can be expressed as

$$\sigma = \alpha_1 \frac{M}{bh_0^2}; \quad \tau = \alpha_2 \frac{V}{bh_0}$$

So

$$\frac{\sigma}{\tau} = \frac{\alpha_1}{\alpha_2} \frac{M}{Vh_0} = \frac{\alpha_1}{\alpha_2} \lambda \tag{4-5}$$

Where α_1, α_2 ——the coefficients related to the form of the beam support and the position of the section;

λ ——generalized shear-span ratio, $\lambda = \frac{M}{Vh_0}$, M is the bending moment of section, and V is the design value of shear force.

λ reflects the relative ratio of σ and τ in the section. To some extent, λ also reflects the relative ratio of bending moment and shear force. It is a decisive factor on the shear failure modes of oblique section of beams without web reinforcements, showing significant influence on the shear load-carrying capacity of the oblique section.

4.2.3 Failure Modes of the Oblique Section

1. The shear failure modes of the oblique section of beam without web reinforcement

Figure 4-6 (a), (b) and (c) respectively show the distribution of the principal stress traces at λ equal to 1/2, 1 and 2. The solid line is the principal tensile stress trace and the dotted line is the principal compressive stress trace.

For different shear-span ratios λ, beams without web reinforcement will fail in different modes.

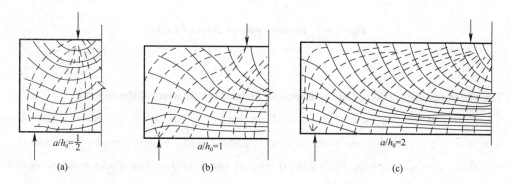

Figure 4-6 The trace of the principal stress

(1) Inclined compression failure ($\lambda < 1$)

If $\lambda < 1$, the concrete between the supports and the nearest concentrated load points is similar to that of an inclined short column. When the concrete is broken, the structure is split into a number of oblique short columns by the web-shear cracks. Therefore, the shear load-carrying capacity depends on the compressive strength of concrete. This mode of failure is called inclined compression failure as shown in Figure 4-7 (a).

(2) Shear-compression failure ($1 \leqslant \lambda \leqslant 3$)

If $1 \leqslant \lambda \leqslant 3$, the damage characteristic is that some vertical cracks appear at the edge of the tensile zone in the flexure-shear sections. The oblique cracks are formed extending along the vertical direction and then extending along the oblique direction, among which a wide inclined crack through the cross section is called the critical oblique crack. It reduces the height of the shear-compressive zone and eventually causes the damages of concrete and loss of load-carrying capacity of the oblique section. This mode of failure is called shear-compression failure as shown in Figure 4-7 (b).

(3) Inclined splitting failure ($\lambda > 3$)

If $\lambda > 3$, the principal tensile stress in the compressive zone exceeds the tensile strength of concrete when the shear stress is rapidly increased in the residual concrete. The inclined crack extends rapidly upward and splits the member into two parts. The splitting plane is clean without much debris. The entire process occurs suddenly and rapidly. The ultimate strength is equal to or slightly higher than the inclined cracking strength. This mode of failure is called inclined splitting failure as shown in Figure 4-7 (c).

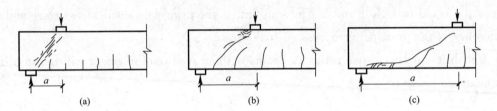

Figure 4-7 Failure modes of diagonal section
(a) Inclined compression failure; (b) Shear-compression failure; (c) Inclined splitting failure

2. The shear failure modes of oblique section of beam with web reinforcement

For the web-reinforcement beams with stirrups, the shear failure mode of the oblique section is based on the beams without web reinforcement, which can also be divided into three failure modes, namely, the inclined compression failure, the shear-compression failure and the inclined splitting failure. At the same time, the number of stirrups has great influence on the failure mode.

When $\lambda > 3$ and the number of stirrups is too small, once the oblique cracks appear, the stirrups immediately yield and cannot control the development of oblique cracks, leading to the inclined splitting failure.

If $\lambda > 3$ and the number of stirrups are appropriate, shear-compression failure occurs. Stirrups can restrain the development of inclined cracks after the formation of oblique cracks and avoid the inclined compression failure. After the stirrups yield, the inclined cracks develop rapidly, reducing the remaining cross section area at the upper end of the inclined fracture and increasing the stress of σ and τ in shear zone of concrete under the joint action to shear-compression failure.

If the number of stirrups is too large, even in the thin web beam or the shear spam λ is large, inclined compression failure will still occur.

4.3 Mechanism of Shear Resistance for Beams without Web Reinforcement

The inclined cracks and vertical cracks separate the beam into two interconnected parts. The upper part is analogous to a two-hinge arch of variable cross sections; the longitudinal bars perform as a tension rod; the arch and the beam have the same supports. The lower part is further divided by cracks into a series of comb-teeth, each of which can be considered as a cantilever beam with the end connected to the inner surface, as shown in Figure 4-8. The root of the tooth is connected to the arch, which is equivalent to the fixed end of a cantilever beam. The force of the tooth is shown in Figure 4-9.

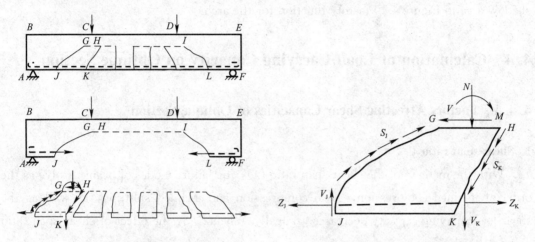

Figure 4-8 The comb-like structure   Figure 4-9 The force of the tooth

The forces acting on the teeth are as follows: 1) the tensile force of longitudinal reinforcement: Z_J and Z_K, $Z_K > Z_J$. 2) the dowel force of longitudinal reinforcement: V_J and V_K. The longitudinal reinforcement is subjected to shear force because there are relative displacements for the concrete on both sides of the crack. 3) the aggregate interlock force between cracks: S_J and S_K. These forces cause the flexure moment M, the axial force N and the shear force V at the root of the comb tooth. M and V are mainly balanced with tension differences of longitudinal and dowel force, and N is mainly balanced with the aggregate interlock force. With the gradual widening of the inclined cracks, the aggregate interlock force decreases, the concrete along the longitudinal reinforcement may be split, and the dowel force of the reinforcement will gradually decrease. At the same time, the behavior of the teeth of comb will be reduced accordingly. The force applied on the arch is shown in Figure 4-10.

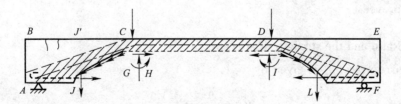

Figure 4-10 The force of the arch

The dash dot is the pressure line of the arch in Figure 4-10. Most sections in the zone of the shear span are basically in a large eccentric compression state. Obviously, the compressive strain at the pressure line is largest. And the farther away from the pressure lines, the smaller the compressive strain will be. Tensile stress may appear at the top. When the load is large, the tensile cracks will appear at the top of the beam. Therefore, zones in the shadow area in Figure 4-10 mainly function for the arch.

4.4 Calculation of Load-Carrying Capacity on Oblique Section

4.4.1 Factors Affecting Shear Capacities of Oblique Section

1. Shear-span ratio

With the increase of the shear-span ratio (λ), the failure modes of beams evolve as the order of inclined compression, shear-compression and inclined splitting, as well as the shear load-carrying capacity decreases gradually. When $\lambda > 3$, the effect of shear span ratio will not be obvious.

2. Strength of concrete

The failure of the oblique section occurs when the concrete reaches the ultimate strength. Therefore, the strength of the concrete has great influence on the shear load-carrying capacity of the beam. When the beam is damaged by inclined compression, the shear load-carrying capacity depends on the compressive strength of concrete. When the beam is damaged by inclined splitting, it depends on the tensile strength of concrete. When the beam is damaged by shear-compression, the effect of concrete strength is between the above two conditions.

3. Longitudinal reinforcement ratio

The shear reinforcement produces the dowel force, which restrains the extension of the diagonal fracture, thereby increasing the height of the shear zone. Therefore, with the increase of the longitudinal reinforcement ratio, the shear load-carrying capacity of the beam is improved.

4. Stirrup Ratio and the strength of stirrups

The stirrup ratio ρ_{sv} is given by

$$\rho_{sv} = \frac{A_{sv}}{bs} = \frac{nA_{sv1}}{bs} \tag{4-6}$$

Where A_{sv}——total area of stirrup within a cross-section of a beam;

s——space of stirrups in the direction of longitudinal reinforcements;

b——width of a rectangular section or the web width of a T-section or an I-section;

A_{sv1}——area of one leg of stirrups;

n——number of stirrups' legs, as shown in Figure 4-11.

In Figure 4-12, the horizontal axis is the stirrup reinforcement ratio (ρ_{sv}) multipling experimental value of stirrup tensile strength (f_{yv}^0). The coordinate $\left(\dfrac{V_u^0}{bh_0}\right)$ is called the experimental value of nominal shear stress. It can be seen from Figure 4-12, with the increases of the stirrups ratio, the shear load-carrying capacity of the oblique section is improved with a linear relationship.

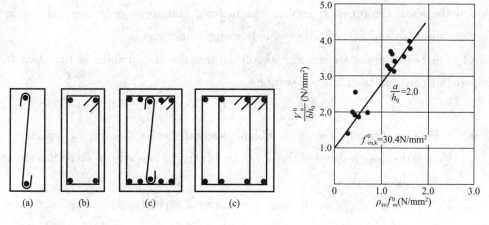

Figure 4-11   Number of stirrups
(a) Single arm hoop; (b) Double arm hoop;
(c) Three arm hoop; (d) Limbs hoop

Figure 4-12   Shear resistance against the value of $\rho_{sv} f_{sv}^0$

5. The aggregate interlock force in oblique section

The aggregate interlock force at the oblique section shows great influence on the capacity of the beams without web reinforcement.

6. Sectional size and shape

(1) The influence of sectional size

The sectional size has great impacts on the shear load-carrying capacity of the beams without web. Tests have shown that when other parameters are constant, the beam height increased by 4 times can lead to the decrease of the average shear stress by 25% to 30%.

(2) The influence of sectional shape

For the T-shaped beam, the size of the flange has influence on the shear load-carrying capacity. If the width of the flange is increased, the capacity can be increased by 25%, but if the flange is too large, the effect tends to be weakened. In addition, with the increase of beam width, the shear capacity is improved.

4.4.2 Design Formulas

1. Basic assumptions

Based on the analysis of various mechanisms of failure, many scholars have proposed various types of formulas for calculating the shear load-carrying capacity of the reinforced concrete beams. However, it is not practical because of the complexity of the methods. The current approach adopted by the Chinese code is semi-theoretical and semi-empirical formulas. These equations ensure a relatively safe state.

The calculation about strength of the beam under shear force is based on the failure mode of the shear-compression failure. The inclined compression failure and the inclined splitting failure should be avoided by the detailing requirements.

(1) For beams under shear-compression failure, the design value of the shear force is composed of three parts in oblique section:

$$\sum Y = 0, \quad V_u = V_c + V_s + V_{sb} \tag{4-7}$$

Where V_u——the design value of the oblique section shear load-carrying capacity;

V_c——the design value of the shear load-carrying capacity in shear-flexure zone of the beam;

V_s——the design value of shear load-carrying capacity of stirrups intersected with inclined cracks;

V_{sb}——the design value of shear load-carrying capacity of bent-up bars intersected with inclined cracks.

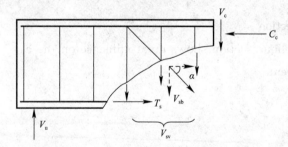

Figure 4-13 The separated diagram for shear capacity calculation

Figure 4-13 shows the diagram for each component.

(2) The aggregate interlock force at the oblique section and the dowel force of the longitudinal reinforcement play important roles in the beams without webs, of which the shear forces occupy 50% to 90% of the total value. However, the test shows the shear force of the above two parts is only about 20% of total value in the beams with webs. In order to simplify the calculation, the contributions of the aggregate interlock force and the dowel force are neglected.

(3) The sectional size mainly affects flexural members without webs. Thus, it is only

considered when calculating the thick plate without stirrups or bent-up bars.

(4) The shear-span ratio is one of the important factors affecting the load-carrying capacity of the oblique section. However, for simpler application of the formula, the effect of λ is considered only when calculating the independent beam based on the concentrated load.

2. Equations for shear capacity of oblique section without web reinforcements

The formulas for calculating the oblique section shear load-carrying capacity of flexural members are mainly based on the experimental results of beams without webs. The test results were divided into two cases:

The first case: The V_c is based on the experimental data of a large number of beams (including simply supported shallow beams without web reinforcement, simply supported beams, simply supported deep beams and continuous shallow beams) under distributed loads and the shear force of the supports (V_u) is taken as the shear load-carrying capacity of the concrete shear zone, as shown in Figure 4-14 (a).

The second case: The V_c is based on the large number of the beams (independent shallow beams under concentrated load, the independent short beams and the independent deep beams), as shown in Figure 4-14 (b).

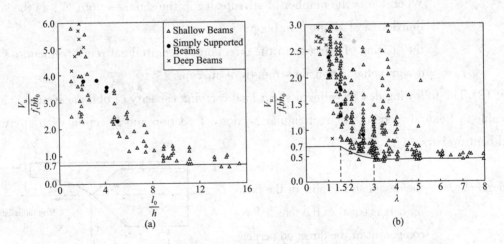

Figure 4-14 Experimental data for shear capacity

(a) Distributed load; (b) Concentrated load

Distributed load
$$V_c = 0.7 f_t b h_0 \tag{4-8}$$

Concentrated load
$$V_c = \frac{1.75}{\lambda + 1} f_t b h_0 \tag{4-9}$$

3. Calculation formula

(1) The following design value of shear load-carrying capacity of oblique section is only applicable to flexural members of rectangular section, T-section and I-section with stirrups.

$$V_u = V_{cs} \tag{4-10}$$

$$V_{cs} = \alpha_{cv} f_t b h_0 + f_{yv} \frac{A_{sv}}{s} h_0 \tag{4-11}$$

Where V_{cs}——the design value of shear load-carrying capacity of concrete and stirrups on oblique section of members;

α_{cv}——coefficient of shear bearing capacity on inclined section, for a generally flexural member, α_{cv} equals to 0.7; for an independent beam subjected to concentrated loads, α_{cv} equals to $\frac{1.75}{\lambda+1}$, the λ is the shear-span ratio of the calculated cross section, which equals to $\frac{a}{h_0}$; When $\lambda < 1.5$, take $\lambda = 1.5$; when $\lambda > 3$, take $\lambda = 3$;

A_{sv}——$A_{sv} = n A_{sv1}$ is the whole sectional area of stirrups leg in the same section; Where, n is the number of stirrups leg in the same section; A_{sv1} is the sectional area of single arm hoop;

s——the spacing of stirrups in the direction of longitudinal reinforcements;

f_{yv}——design value of tensile strength of stirrups.

(2) The following design value of shear load-carrying capacity of oblique section is applicable to flexural members of rectangular section, T-section and I-section with stirrups and bent-up bars.

$$V_u = V_{cs} + V_{sb} \tag{4-12}$$

Where V_{sb}——the design shear value of the bent-up bars is equal to the tensile force component in the direction perpendicular to the beam axis, as is shown in Figure 4-15, and can be calculated according to the following formula:

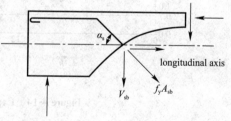

Figure 4-15 The shear force of the bent-up bars

$$V_{sb} = 0.8 f_y A_{sb} \sin\alpha_s \tag{4-13}$$

Therefore

$$V_u = \alpha_{cv} f_t b h_0 + f_{yv} \frac{A_{sv}}{s} h_0 + 0.8 f_y A_{sb} \sin\alpha_s \tag{4-14}$$

Where f_y——the design value of tensile strength of bent-up bars;
A_{sb}——the area of bent-up bars in the same section;
α_s——the angle between the bent-up bars in oblique section and the longitudinal axis of the member, which is 45°. When the beam section exceeds 800mm, it is usually taken as 60°.

The coefficient 0.8 in the formula is the reduction for the shear capacity of bent steel bars. This is due to the fact that the bent-up bars intersected with the oblique crack may approach the shear compression zone and cannot reach yield strength when the oblique section is subjected to shear failure.

(3) The design value of shear load-carrying capacity of inclined section for general flexural members without stirrups or bent-up bars is obtained as:

$$V_u = 0.7\beta_h f_t b h_0 \tag{4-15}$$

$$\beta_h = \left(\frac{800}{h_0}\right)^{\frac{1}{4}} \tag{4-16}$$

Where β_h——coefficient of section depth; if $h_0 < 800$mm, take $h_0 = 800$mm; If $h_0 \geqslant 2000$mm, take $h_0 = 2000$mm.

4. The description of the formulas

(1) V_{cs} consists of two items. The first $(\alpha_{cv} f_t b h_0)$ is the shear force obtained from the shear zone of the concrete. And the second $\left(f_{yv}\dfrac{A_{sv}}{s}h_0\right)$ is the shear force mostly obtained from the stirrups with only a small part from concrete. Both the stirrups and shear compression zone contribute to the shear load-carrying capacity. However, it is difficult to separate one from another, and it is not necessary. Therefore, V_{cs} should be considered as the shear force combination of concrete in shear-compression zone and stirrups.

(2) It is assumed that the horizontal projection length of the oblique shear crack is h_0, and the stirrup can reach the tensile design strength, so the shear force is $f_{yv}\dfrac{A_{sv}}{s}h_0$.

(3) $\alpha_{av} = 0.7 \sim 0.44$ is corresponding to $\lambda = 1.5 \sim 3.0$. It is shown that the shear load-carrying capacity of the independent beams without web reinforcement under uniform load is lower than the other beams when $\lambda > 1.5$. The higher the value λ is, the larger the decrease will be.

(4) To be specified: concrete beams should use stirrups as shear reinforcement, taking into account the convenience of design and construction. At present, the general beam (except the cantilever beam) and the plate in the construction engineering do not utilize bent-up bars any more. Whereas in the bridge engineering, bent-up bars is still the common style.

(5) The Eqs. (4-11) and (4-14) are applicable to rectangular, T-section and I-section members. It has shown that the section shape can affect the bearing capacity, but the effect can be ignored.

5. Applicable range of formula

(1) The minimum size of the cross section (Upper bound of shear capacity)

Firstly, in order to avoid the inclined compression failure, the sectional size of beam should not be too small, which is the main reason. Secondly, in order to prevent the over-width of the crack at the service stage (mainly to the thin web beam), it is recommended in GB 50010 that the ultimate shear resistance of diagonal section should satisfy the following requirements:

For

$$\frac{h_w}{b} \leqslant 4, \ V \leqslant 0.25\beta_c f_c b h_0 \tag{4-17}$$

For

$$\frac{h_w}{b} \leqslant 6, \ V \leqslant 0.2\beta_c f_c b h_0 \tag{4-18}$$

When $4 < \frac{h_w}{b} < 6$, the upper bound value is determined by interpolation.

Where V——design value of shear force;

β_c——influence coefficient of concrete strength, if the grade of concrete is less than or equal to C50, $\beta_c = 1.0$; if the grade of concrete is equal to C80, $\beta_c = 0.8$; the grade within the range is determined by interpolation;

f_c——design value of compressive strength of concrete;

b——the width of the rectangular section, for the T or I section, b is the web width;

h_w——$h_w = h_0$ to a rectangular section; h_w is equal to the depth of web for a T-section and an I-section.

The upper bound of shear resistance is dependent on the dimensions of a section and the compressive strength of concrete, but is independent of the amount of web reinforcement. The recommended limiting values for the ultimate diagonal section resistance enable to restrict the width of diagonal cracks.

(2) Minimum quantities of stirrups (Lower bound of shear capacity)

If the number of stirrups is too small, once the oblique cracks appear, the tensile stress in the stirrups may suddenly increase to reach the yield strength, resulting in the acceleration of the crack, and even the fracture of the stirrup, which will finally result in the inclined splitting failure. In order to avoid this failure, for $V > 0.7 f_t b h_0$, the stirrup reinforcement ratio ρ_{sv} should not be less than the minimum reinforcement ratio of $\rho_{sv, min}$:

$$\rho_{sv} = \frac{A_{sv}}{bs} \geqslant \rho_{sv, min}$$

$$\rho_{sv, min} = 0.24 \frac{f_t}{f_{yv}}$$

(4-19)

4.4.3 Calculation Method of Shear Load-Carrying Capacity of Oblique Section

1. The positions of the cross sections for calculation

(1) Section at the edge of supports (section 1-1) in Figure 4-16.

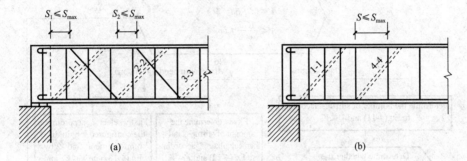

Figure 4-16 Spacing of web reinforcement and the positions of the critical sections
(a) Beam with both stirrups and bent-up bars; (b) Beam with stirrups only

(2) Section at the bent-up point of bent-up bars (section 2-2 and 3-3 in Figure 4-16);

(3) Section at the position where there is a change in the spacing of stirrups (section 4-4 in Figure 4-16).

2. Calculation procedures

The calculation of bearing capacity of reinforced concrete beam includes calculation of flexural capacity of normal section and calculation of shear capacity of oblique section. Usually the latter is based on the former, that is, cross-sectional size and longitudinal reinforcement have been initially selected. At this point, the cross-sectional size of the component can be checked by the upper limit of oblique sectional shear capacity to avoid the occurrence of oblique-compression failure. If it is not satisfied, the cross-sectional size should be readjusted. And then the oblique-sectional shear capacity can be calculated by formulas. The appropriate stirrups and bent-up bars can be configured according to the calculated results. The reinforcement ratio of stirrups should meet the requirements of the minimum reinforcement ratio to avoid the inclined splitting failure. When the conditions are under $0.7 f_t b h_0 \geqslant V$ or $\frac{1.75}{\lambda+1} f_t b h_0 \geqslant V$, stirrups can be configured according to structural requirements.

The block diagram for calculation of oblique sectional shear capacity for reinforced concrete beam is shown in Figure 4-17.

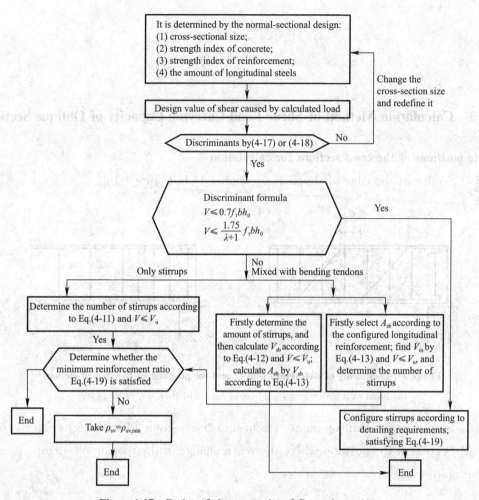

Figure 4-17 Design of shear capacity of flexural members

4.4.4 Examples

[**Example 4-1**] Figure 4-18 shows a simply supported reinforced concrete beam with a clear span $l_n = 3.56$m. The section of the beam is rectangular with $b \times h = 200\text{mm} \times 500\text{mm}$. The concrete is of grade C30. The longitudinal reinforcement is of HRB335 and the transverse stirrup is of HPB300. The characteristic value of the uniformly distributed permanent load is $g_k = 25$kN/m, and the characteristic value of the uniformly distributed variable load is $q_k = 50$kN/m. It is required to design the web reinforcements of the beam.

[**Data**] C30 concrete: $f_t = 1.43\text{N/mm}^2$, $f_c = 14.3\text{N/mm}^2$

HRB335 steel: $f_y = 300\text{N/mm}^2$

HPB300 steel: $f_{yv} = 270\text{N/mm}^2$

[**Solution**]

(1) Determine the critical section and calculate the design value of shear force V.

The critical section is at the edge of supports, the design value of shear force V is:

$$V = \frac{1}{2}(\gamma_G g_k + \gamma_Q q_k)l_n = \frac{1}{2}(1.2 \times 25 + 1.4 \times 50) \times 3.56 = 178 \text{kN}$$

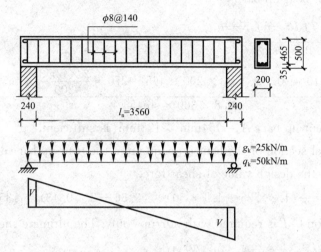

Figure 4-18 Beam with stirrups only (unit: mm)

(2) Check for the adequacy of concrete section for shear:

$$h_w = h_0 = 465 \text{mm}, \quad \frac{h_w}{b} = \frac{465}{200} < 4, \quad \beta_c = 1.0$$

$$0.25\beta_c f_c b h_0 = 0.25 \times 1.0 \times 1.43 \times 200 \times 465$$
$$= 332475\text{N} > V = 178\text{kN}$$

Hence the concrete section is adequate.

(3) Check whether web reinforcements need to be calculated

$$0.7 f_t b h_0 = 0.7 \times 1.43 \times 200 \times 465$$
$$= 93093\text{N} = 93.093\text{kN} < V = 178\text{kN}$$

Hence the web reinforcements should be decided by calculation.

(4) Calculate the web reinforcements

1) With stirrups only

$$\frac{nA_{sv1}}{s} \geq \frac{V - 0.7 f_t b h_0}{f_{yv} h_0} = \frac{178 \times 10^3 - 0.7 \times 1.43 \times 200 \times 465}{270 \times 465}$$
$$= 0.676 \text{mm}^2/\text{mm}$$

Take two-legged $\phi 8@140$ stirrups: $A_{sv1} = 50.3\text{mm}^2$, $n=2$.

From the above equation, the value $s \leq 148.8$mm, and $s = 140$mm is determined.

Then $\rho_{sv} = \dfrac{nA_{sv1}}{bs} = \dfrac{2 \times 50.3}{200 \times 140} = 0.359\% > \rho_{sv,\min} = 0.24 \dfrac{f_t}{f_{yv}} = 0.127\%$

So the stirrup ratio is sufficient.

2) With both stirrups and bent-up bars

The configuration of two-legged $\phi 8$ stirrups is $A_{sv1} = 50.3\text{mm}^2$, $n = 2$, and the spacing of stirrups is $s=200$mm according to the constructional requirements. Assume that the angle α of bent up is 45°. At the surface of the support (diagonal section A-I in Figure 4-19), the required area of bent-up bar is

$$A_{sb} = \frac{V - 0.7 f_t b h_0 - f_{yv} \frac{A_{sv}}{s} h_0}{0.8 f_y \sin\alpha}$$

$$= \frac{178 \times 10^3 - 0.7 \times 1.43 \times 200 \times 465 - 270 \times \frac{2 \times 50.3}{200} \times 465}{0.8 \times 300 \times \sin 45°} = 128 \text{mm}^2$$

Take 1 Φ 14 bent-up bar, $A_b = 154 \text{mm}^2 > 128 \text{mm}^2$ is sufficient.

At the diagonal section of the bent-up point from the bent-up bar (diagonal section C-J in Figure 4-19), the design value of shear force

$$V_1 = \frac{1}{2}(1.2 \times 25 + 1.4 \times 50) \times (3.56 - 2 \times 0.43) = 135 \text{kN}$$

Diagonal section C-J is required with stirrups only. The ultimate shear strength is

$$V_{cs} = 0.7 f_t b h_0 + f_{yv} \frac{n A_{sv1}}{s} h_0$$

$$= 0.7 \times 1.43 \times 200 \times 465 + 270 \times \frac{2 \times 50.3}{200} \times 465$$

$$= 156244 \text{N} > V_1 = 135 \text{kN}$$

The web reinforcement is sufficient. The stirrups and the bent-up bars are shown in Figure 4-19.

[**Example 4-2**] Figure 4-20 shows a simply hinged beam under a concentrated load, the design value is $P=700$kN (self weight of the beam is neglected). The section of the beam is rectangular of $b \times h = 250\text{mm} \times 700\text{mm}$ with a span $l=4.0$m. The concrete is of grade C30. Both the stirrups and the bent-up bars are required.

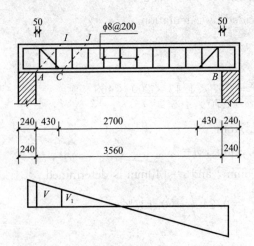

Figure 4-19 Beam with both stirrups and bent-up bars (unit: mm)

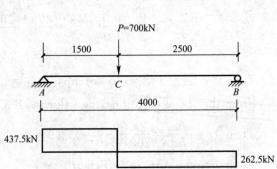

Figure 4-20 Simply supported beam under concentrated load (unit: mm)

[**Data**]   C30 concrete: $f_t = 1.43 \text{N/mm}^2$, $f_c = 14.3 \text{N/mm}^2$

HRB400 steel: $f_y = 360 \text{N/mm}^2$

$h_0 = h - 35 = 665 \text{mm}$

[**Solution**]

The calculation procedure is listed in Table 4-1.

Calculation procedure   Table 4-1

Position of the critical sections	The left section of point "C"	The right section of point "C"
Design value of shear force V	437.5kN	262.5kN
$\lambda = \dfrac{a}{h_0}$	1.5<2.26<3 Take $\lambda = 2.26$	3.76>3 Take $\lambda = 3$
$0.25\beta_c f_c b h_0$	594.344kN>V=437.5kN The concrete section is sufficient	594.344kN>V=262.5kN The concrete section is sufficient
$\dfrac{1.75}{\lambda+1.0} f_t b h_0$	127.62kN<V=437.5kN The web reinforcement is determined by calculations	104.01kN<V=262.5kN The web reinforcement is determined by calculations
Stirrup	Two-legged $\phi 8$ stirrups, $s=200$mm	Two-legged $\phi 8$ stirrups, $s=200$mm
$v_{cs} = \dfrac{1.75}{\lambda+1.0} f_t b h_0 + f_{yv} \dfrac{nA_{sv1}}{s} h_0$	248.04kN	224.43kN
$A_{sb} = \dfrac{V - V_{cs}}{0.8 f_y \sin\alpha}$	930.3mm² ($\alpha = 45°$)	187mm² ($\alpha = 45°$)
Bent-up bars	2⚛25, A_{sb}=982mm²	1⚛25, A_{sb}=491mm²

[**Example 4-3**]   A Simply hinged reinforced concrete beam with a span $c = 4.0$m is under both concentrated loads and uniformly distributed loads. The design values of loads are shown in Figure 4-21. The section of the beam is rectangular with $b \times h = 200$mm $\times$ 600mm. The concrete is of grade C30 and the transverse stirrups are of HRB335. It is required to design the stirrups of the beam.

[**Data**]   C30 concrete: $f_t = 1.43 \text{N/mm}^2$, $f_c = 14.3 \text{N/mm}^2$

HRB400: $f_y = 360 \text{N/mm}^2$

$h_0 = h - 35 = 565 \text{mm}$

[**Solution**]

(1) Determine the critical section and calculate the design value of shear force

The beam is divided into four segments of AC, CD, DE and EB, in which the bending moments and the shear forces are shown in Figure 4-21. Because the beam is symmetrically loaded, only the two segments AC and CD are calculated.

(2) Segments AC and EB

The beam is under both a concentrated load and distributed loads. The shear force by concentrated load is $V_1 = 180$kN and the total shear force is

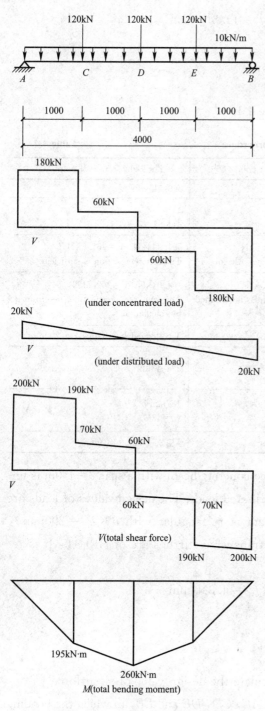

Figure 4-21 Beam under both concreted load and distributed load (unit: mm)

$V_{1\max} = 200\text{kN}, \dfrac{V_1}{V_{1\max}} = \dfrac{180}{200} = 90\% > 75\%$

The shear span ratio is calculated as:

$\lambda = \dfrac{M}{Vh_0} = \dfrac{195 \times 10^6}{190 \times 10^3 \times 565} = 1.816$

And $1.5 < 1.816 < 3$, take $\lambda = 1.816$

$0.25\beta_c f_c bh_0 = 0.25 \times 1.0 \times 14.3 \times 200 \times 565$
$= 403975\text{N} = 403.975\text{kN} > V_{1\max}$
$= 200\text{kN}$

Thus, the concrete section is sufficient:

$\dfrac{1.75}{\lambda + 1.0} f_t bh_0 = \dfrac{1.75}{1.816 + 1.0} \times 1.43 \times 200 \times 565$
$= 100420\text{N} = 100.42\text{kN} < V_{1\max}$
$= 200\text{kN}$

Thus, the web reinforcements should be decided by calculations:

$\dfrac{nA_{sv1}}{s} \geqslant \dfrac{V_{1\max} - \dfrac{1.75}{\lambda + 1.0} f_t bh_0}{f_{yv} h_0}$

$= \dfrac{200 \times 10^3 - \dfrac{1.75}{1.816 + 1.0} \times 1.43 \times 200 \times 565}{300 \times 565}$

$= 0.587 \text{mm}^2/\text{mm}$

Take two-legged ϕ 8 stirrup of $A_{sv1} = 50.3\text{mm}^2$ and $n = 2$.

Based on the above equation, spacing is obtained as $s \leqslant 171\text{mm}$, and $s = 150\text{mm}$ is taken.

Then

$\rho = \dfrac{nA_{sv}}{bs} = \dfrac{2 \times 50.3}{200 \times 150} = 0.335\% > \rho_{sv,\min}$

$= 0.24 \dfrac{f_t}{f_{yv}} = 0.114\%$

Hence the stirrup ratio is sufficient.

(3) Segments CD and DE

The shear force produced by the concentrated load is $V_2 = 60\text{kN}$, and the total shear force is

$$V_{2max} = 70\text{kN} \quad \frac{V_2}{V_{2max}} = \frac{60}{70} = 85.7\% > 75\%$$

Thus, the shear span ratio is calculated as:

$$\lambda = \frac{M}{Vh_0} = \frac{260 \times 10^3}{60 \times 10^3 \times 0.565} = 7.67 > 3$$

Take $\lambda = 3$

$$\frac{1.75}{\lambda + 1.0} f_t b h_0 = \frac{1.75}{3 + 1.0} \times 1.43 \times 200 \times 565 = 70696\text{N}$$
$$= 70.696\text{kN} > V_{2max} = 70\text{kN}$$

Thus, the required web reinforcements can be set by constructional requirements. Take two-legged $\Phi 8$ stirrups of spacing at $s = 250\text{mm}$.

[**Example 4-4**] A simply supported reinforced concrete beam under uniformly distributed load q is given. The section of the beam is rectangular with $b \times h = 200\text{mm} \times 400\text{mm}$ of a clear span $l_n = 4.26\text{m}$. The concrete is of grade C20 and the transverse stirrups are of HPB300. Two-legged $\Phi 8$ stirrups are provided at a spacing of $s = 200\text{mm}$. The maximum value of the distributed load q carried by the beam is required according to the requirements of Chinese Design code.

[**Data**] C20 concrete: $f_t = 1.1\text{N/mm}^2$, $f_c = 9.6\text{N/mm}^2$

HPB300 steel: $f_y = 270\text{N/mm}^2$, $\phi 8$ stirrup: $A_{sv1} = 50.3\text{mm}^2$

$h_0 = h - 35 = 365\text{mm}$

[**Solution**]

(1) Calculate V_{cs}

$$V_{cs} = 0.7 f_t b h_0 + f_{yv} \frac{A_{sv}}{s} h_0 = 0.7 \times 1.1 \times 200 \times 365 + 270 \times \frac{2 \times 50.3}{200} \times 365$$
$$= 105780\text{N} = 105.78\text{kN}$$

(2) Check for the adequacy of concrete section and the stirrup ratio

$$0.25\beta_c f_c b h_0 = 0.25 \times 1.0 \times 9.6 \times 200 \times 365$$
$$= 175200\text{N} = 175.2\text{kN} > V_{cs} = 105.78\text{kN}$$

$$\rho_{sv} = \frac{nA_{sv1}}{bs} = \frac{2 \times 50.3}{200 \times 200} = 0.25\% > \rho_{sv,\min} = 0.24 \frac{f_t}{f_{yv}} = 0.098\%$$

Hence the concrete section and stirrup ratio are sufficient, and the ultimate shear strength is:

$$V_u = V_{cs} = 105.78\text{kN}$$

(3) Calculate the value of the distributed load q

$$q = \frac{2V_u}{l_n} = \frac{2 \times 105.78}{4.26} = 49.66\text{kN/m}$$

4.5 Measures to Ensure the Flexural Capacities of Inclined Sections in Flexural Members

As mentioned in Section 4.1, the bearing capacity of the oblique section includes the shear capacity and the bending capacity. Oblique sectional flexural capacity of the beam refers to the sum of internal moments caused by the tensions multiplying the distances to the shear-compression zone A, including the longitudinal tensile bars, bending bars, stirrups and so on when the oblique section is damaged ($M_u = F_s \cdot z + F_{sv} \cdot z_{sv} + F_{sb} \cdot z_{sb}$), as shown in Fingure 4-22.

However, the flexural capacity of the oblique section is usually not calculated, but ensured by bending, truncation, anchorage of longitudinal reinforcement, stirrups spacing and other structural measures. Therefore, based on the calculation and structure of the flexural capacity of a single normal section in the last chapter, this section will emphasize the reinforced structure and other issues along the entire length of bending member.

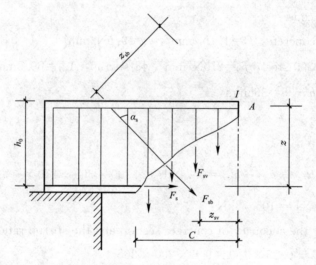

Figure 4-22 Calculation of oblique sectional flexural capacity of bending member

4.5.1 Moment Capacity Diagram

The diagram plotted by the moment design value M produced by each normal section of beam under load is called the moment diagram, namely the M diagram. The diagram plotted by design value of flexural capacity M_u from each normal section of beam according to co-working of the reinforcement and concrete is called the bending capacity diagram of normal section. Because M_u is provided by the properties of materials, M_u diagram is also known as the material diagram.

In order to meet the requirement of $M_u \geqslant M$, M diagram must be wrapped in M_u diagram to ensure the flexural capacity of each normal section in a beam.

Figure 4-23 shows the reinforcement diagram, M diagram and M_u diagram of a simply supported beam with a uniform load. The beam is configured with longitudinal reinforcements of $2 \oplus 22 + 1 \oplus 20$. If the area of the longitudinal reinforcement is equal to the calculated area required for the mid-span section, the peripheral horizontal line of M_u diagram is exactly tangent to the maximum moment point of the M diagram. If the total area of the longitudinal reinforcement is slightly larger than the calculated area, the peripheral horizontal line of M_u diagram can be obtained by the following Eqs. (4-20) and (4-21) according to the actual amount of reinforcement A_s, which is:

$$M_u = A_s f_y \left(h_0 - \frac{f_y A_s}{2\alpha_1 f_c b} \right) \tag{4-20}$$

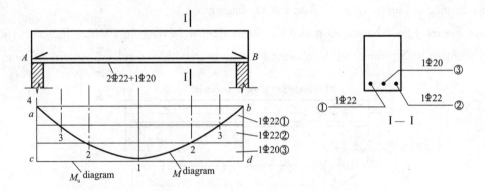

Figure 4-23 **Flexural capacity diagram of normal section for simply supported beam with entire long longitudinal reinforcement**

The flexural capacity M_{ui} provided by any longitudinal tensioned bar can be approximately obtained by the ratio of sectional area A_{si} to the total sectional area A_s of the steel bars, multiplied by M_u, which is:

$$M_{ui} = \frac{A_{si}}{A_s} M_u \tag{4-21}$$

For example in Figure 4-23, if both ends of the three bars are assembled into the support, the M_u diagram is the total area of $acdb$. The moment M_{ui} provided by every steel is marked on the figure with horizontal lines. In practical design, usually certain longitudinal reinforcements are bent up, taking use of their shear performance for effective comsumption. Because the longitudinal tensile reinforcement at the bottom of the beam can't be cut off and the number putting into the support can't be less than two, the bending steel is only ③ of $1 \oplus 20$, which is plotted on the outside for M_u diagram.

From Figure 4-23, ③ steel is fully utilized in the section 1; ② steel is fully utilized in the section 2; ① is fully utilized in the section 3. Therefore, the cross-sections 1, 2 and 3 can be called the section with full use of ③, ②, ① steel respectively. From Figure 4-23, it can be known that the steel ③ is redundant in section 2, same with ② in section 3 and ① in section 4, Therefore, sections 2, 3 and 4 can be called no-need sections of ③, ②, ① steel respectively.

If ③ steel is bent up near the support, the bending point e and f must be outside the section 2 (Figure 4-24). It can be approximately considered that when the bent bar is at the centerline of the sectional height with no contributions to the flexural capacity, the M_u diagram is the total area of $aigefhjb$ in Figure 4-24. In the figure, the points e and f vertically correspond to the intersection points E and F, respectively. In the same way, the points g and h vertically correspond to the intersection points G and H, respectively. As the bending inner-force arm of normal section for bent-up steel gradually reduces, the flexural capacity of normal section reduces accordingly, leading to the diagonal lines, eg and fh on the M_u diagram.

In Figure 4-24, the points g and h fall outside the M diagram, which indicates the M_u diagram after longitudinal reinforcement bent up is still able to fully cover the M diagram.

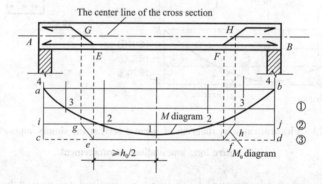

Figure 4-24 The flexural capacity diagram of normal section of a simply supported beam with bent steel bars

4.5.2 Detailing Requirements to Ensure the Flexural Capacities of Inclined Sections with Bent-Up Bars

The previous methods merely discuss the flexural capacity of normal section for bending steels, the following chapter will describe the distance where the longitudinal tensile steel bars should be bent-up from the fully utilized section to ensure the flexural capacity of oblique section.

1. The position of the bending point

In Figure 4-25, the cross-sectional area of the longitudinal tensile steel bar is A_{sb}. Before bending up, the flexural capacity in the normal section I-I is:

$$M_{u, \text{I}} = f_y A_{sb} z \tag{4-22}$$

After bending up, the flexural capacity in the oblique section II-II is:

$$M_{u, \text{II}} = f_y A_{sb} z_b \tag{4-23}$$

According to Figure 4-25, the design value of bending moment undertaken by the oblique section II-II is the design value of the bending moment undertaken by the normal section I-I at the shear compression zone of the oblique sectional end. That is, in order to ensure the flexural capacity of oblique section, the value should be no less than the capacity of the normal section, namely, $M_{u, \text{I}} = M_{u, \text{II}}$, $z_b = z$.

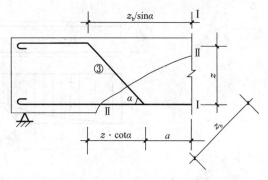

Figure 4-25 The position of bending starting point

Set the distance from the bending up point to the section I-I as a, which is seen from Figure 4-25.

$$\frac{z_b}{\sin\alpha} = z\cot\alpha + a \tag{4-24}$$

$$a = \frac{z_b}{\sin\alpha} - z\cot\alpha = \frac{z(1-\cos\alpha)}{\sin\alpha} \tag{4-25}$$

Usually, take $\alpha = 45°$ or $60°$, and take approximately $z = 0.9h_0$.

Thus

$$a = (0.373 \sim 0.52)h_0 \tag{4-26}$$

For convenience, *Code for Design of Concrete structures* stipulates that the distance between the bending point and the section of full use of the reinforcement shall not be less than $0.5 h_0$. Thus, the distance between the point e and the section 1 should be greater than or equal to $h_0/2$ in Figure 4-24.

For a continuous beam, the rules should be followed both in the mid-span bearing positive moment and at the support bearing the negative moment. For the bars a and b in Figure 4-26, when it is bent up in the positive moment section at the bottom of the beam, the distance between the bending point a and the fully utilized section 4 should be greater than or equal to $h_0/2$. In addition, the distance between the bending point b and the fully utilized section 4 should also be greater than or equal to $h_0/2$ in the negative moment section at the top of the beam. Otherwise, the bent-up bar can't be used as a negative steel on the supported section.

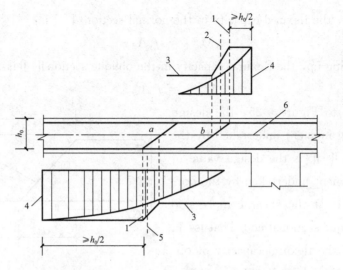

Figure 4-26 The relationship between the bending point and the bending moment diagram
1—the bending section in tension zone; 2—the cross section that the external moments can be supported without reinforcement according to calculation; 3—the flexural capacity without reinforcement; 4—the section of the fully utilized reinforcement a and b according to calculation; 5—the section that the positive moments can be supported without reinforcemant; 6—the center line of beam

2. The end of bending point

As shown in Figure 4-27, the distance between the end of bending point and the end of the support or the start of bending point from the front row of bending bars should not be greater than the maximum spacing of the stirrups. The values are shown in the column of $V>0.7f_t bh_0$ in Table 4-2. This requirement is to make each bent steel intersect with the oblique cracks to ensure the shear and flexural capacity of the oblique section.

3. Anchoring of the bending steel, duck steel and floating steel

The end of the bent-up steels should also leave a certain length for anchoring: the distance in tension zone should not be less than $20d$; the distance should not be less than $10d$ in compression zone; the smooth curved steel bars should also set the hook at the end, as shown in Figure 4-28.

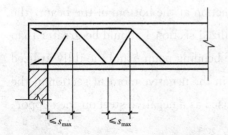

Figure 4-27 The end of bending point

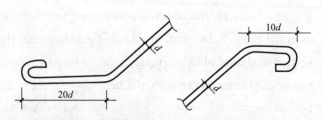

Figure 4-28 The anchor at the end of bending steel

The angle steel at the bottom or top of the beam and the steels at both sides of the beam may not be bent.

Besides bending the longitudinal reinforcement, extra bending bars can be set, known as duck steel shown in Figure 4-29(a). The effect of the duck steel is to transfer the diagonal compression from the concrete among diagonal cracks to the concrete in compression zone, in order to strengthen the co-working between different blocks, by which an anchor truss is formed, the single floating steel shown in Figure 4-29 (b) is forbidden in the design.

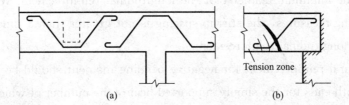

Figure 4-29 Hanging steel, duck steel and floating steel

4.5.3 Anchorage of Longitudinal Reinforcement at the Supports

After the oblique cracks occur at the support of the simply supported beam, the stress of longitudinal reinforcements will increase. At this time, the flexural capacity of beam also depends on the anchoring of the longitudinal reinforcement at the support. If the anchorage length is insufficient, the relative sliding between the reinforcement and the concrete will increase the width of oblique cracks significantly.

The lower longitudinal reinforcement of the simply supported beam and the simply support in the continuous beam should have a certain anchorage length extending into the support. Considering that there is a favorable effect of lateral compressive stress at the support, the anchor length is slightly smaller than the basic anchorage length. According to *Code for Design of Concrete Structures*, the anchorage length l_{as} of the simply supported beam with lower reinforcement ratio extending into the range of support (Figure 4-30) should meet the following conditions:

When $V \leqslant 0.7 f_t b h_0$, $l_{as} \geqslant 5d$; when $V > 0.7 f_t b h_0$, $l_{as} \geqslant 12d$ for ribbed steel bar and $l_{as} \geqslant 15d$ for plain bar, where d is the diameter of the anchorage reinforcement.

When l_{as} can't satisfy these requirements, other effective anchorage measures should be taken, such as hook or mechanical anchoring measures.

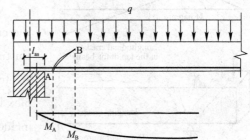

Figure 4-30 Anchor of reinforcement at the support

For the simply supported beam and continuous beam that strength grade of concrete is less than or equal to C25, if there is concentrated load within 1.5h from the bearing and $V>0.7f_tbh_0$, the ribbed steel bar should be taken effective anchorage measures, or taken $l_{as}\geqslant 15d$.

The reinforced concrete independent beam supported in the masonry structure should be equipped with no less than two stirrups within the range of anchorage length l_{as} of longitudinal reinforcement. The diameter of the stirrups should not be less than 0.25 times the maximum diameter of the longitudinal reinforcement, and spacing should not be greater than 10 times the minimum diameter of the longitudinal reinforcement. When taking mechanical anchoring measures, the stirrup spacing is not greater than 5 times the minimum diameter of the longitudinal reinforcement.

The structural reinforcement for negative bending moment should be provided on the upper part at both ends for the simply supported beam, the number of which should not be less than 1/4 of the longitudinal reinforcement and not less than two.

4.5.4 Detail Requirements to Ensure the Flexural Capacities of Inclined Sections When Longitudinal Bars are Cut off

Because of the large range of positive bending diagram of the beam and the tension zone almost covering the entire span, the bottom longitudinal reinforcement of beam should not be cut off. For the longitudinal flexural reinforcement at the top beam within the negative bending moment section near the support, because the range of negative bending moment is not large, some steel bars are cut off to reduce the number of longitudinal bars, but the cut-off point should not be in the tension zone.

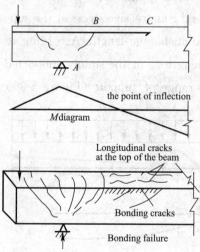

Figure 4-31 The bonding behavior of the cutting-off bars

In the span of the continuous beam and the frame beam, it is necessary to cut off some negative bending moment reinforcement, and the cut-off point should meet the following conditions:

(1) The length from the fully utilized section to the cut-off point is known as the stretched length. Cutting negative steel must meet the requirements of stretched length. This is because the cracks are complex in the negative moment section of the support, including vertical cracks, diagonal cracks and bonding cracks, as shown in Figure 4-31. Sufficient anchoring length should be left for better operating.

(2) The length from the section without the

reinforcement to the cut-off point is called extension length. In order to ensure the bearing capacity of the oblique section, the negative reinforcement must meet the requirement of extension length when cut off.

According to the mentioned reasons, for the longitudinal tensile reinforcement under negative bending moment at the support section of beam cutting off, both the requirements of stretched length and extension length should be met (Figure 4-31), which is divided into three cases.

Case 1: When $V \leqslant 0.7 f_t b h_0$ as shown in Figure 4-32 (a), the stretched length is not less than $1.2 l_a$ and the extension length is not less than $20d$.

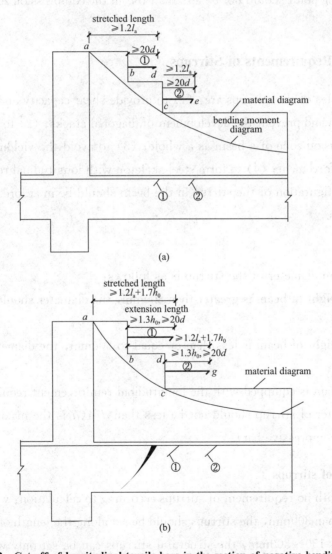

Figure 4-32 Cut-off of longitudinal tensile bars in the section of negative bending moment
(a) $V \leqslant 0.7 f_t b h_0$; (b) $V > 0.7 f_t b h_0$ and the cut-off point is in the tension zone of negative bending moment

Case 2: When $V>0.7f_tbh_0$, the stretched length is not less than $1.2l_a+h_0$ and the extension length is not less than h_0 and $20d$.

Case 3: When $V>0.7f_tbh_0$ and the cutting-off point is still in the tension zone of negative bending moment, the stretched length is not less than $1.2l_a+1.7h_0$ and the extension length is not less than $1.3h_0$ and $20d$.

There should be not less than two upper bars extending to the outer end of the cantilever beam, and the length for bending down vertically is not less than $12d$. The rest of the steels should not be cut off at the upper part of the beam, which should be bent down at the specified point. Anchoring in the bottom of the beam, the anchor length extended from the end of bending point should not be less than $10d$ in the compression zone and $20d$ in the tension zone.

4.5.5 Detail Requirements of Stirrups

The main roles of the stirrups are: (1) to provide shear capacity and flexural capacity of oblique section and prevent the development of diagonal cracks; (2) to connect compression zone and tension zone of a beam as a whole; (3) to avoid the yielding of longitudinal compressive reinforcement; (4) to form steel skeleton with longitudinal reinforcement. The diameter and configuration of the stirrup in the beam should be in accordance with the following provisions.

1. Diameter

The minimum diameter of the stirrup is as follows:

When the height of beam is greater than 800mm, the diameter should not be less than 8mm;

When the height of beam is less than or equal to 800mm, the diameter should not be less than 6mm;

When the beam is equipped with the longitudinal reinforcement required by the calculation, the diameter of stirrup should not be less than $d/4$ (d is the maximum diameter of the longitudinal compressive bar).

2. Configuration of stirrups

For beams with no requirement of stirrups according to calculation: when the height of beam is greater than 300mm, the stirrups should be set along the length of beam; when the height of beam is 150~300mm, the structural stirrups can be set only within the $l_0/4$ of the ends of the component (l_0 is the span of beam). However, when there is a concentrated load in the middle span of the component, stirrups should be set along the length of beam.

when the height of the beam is below 150mm, there is no requirement of stirrups.

The spacing of the stirrups should be determined in accordance with the calculated requirements. In addition, the maximum spacing should also meet the requirements in Table 4-2. When $V>0.7f_tbh_0$, the reinforcement ratio of stirrup should not be less than $0.24f_t/f_{yv}$.

The maximum spacing of the stirrups in the beam(mm)　　　　Table 4-2

The height of beam h	$V>0.7f_tbh_0$	$V\leqslant 0.7f_tbh_0$
$150<h\leqslant 300$	150	200
$300<h\leqslant 500$	200	300
$500<h\leqslant 800$	250	350
$h>800$	300	400

The spacing of the stirrups should not be greater than $15d$ in the colligation of frame and should not be greater than 400mm (d is the smallest diameter among the longitudinal compression bars). This is to ensure that the setting of the stirrups is coordinated with the compression bars to prevent buckling of the compression bars. Therefore, when the beam is equipped with the longitudinal compression bars according to calculation, the stirrups must be made in a closed form, as shown in Figure 4-33(a). When the width of beam is greater than 400mm and the number of longitudinal compression bars in a layer are more than 3, or when the width of beam is not greater than 400mm, but the number of longitudinal compression bars in a layer are more than 4 without setting the composite stirrups, or when the number of longitudinal compression bars in a layer are more than 5 with the diameter greater than 18mm, the spacing of stirrups must be less than or equal to $10d$ (d is the minimum diameter of longitudinal compressive bars).

When the longitudinal reinforcement in a tied frame of the beam is lapped without welding, the stirrups' diameter should not be less than 0.25 times of the diameter of the lap within the length of lap, and the spacing of the stirrups should meet the following requirements:

For tension longitudinal reinforcement, the stirrups' spacing should not be greater than $5d$ and 100mm.

For compressive longitudinal reinforcement, stirrups' spacing should not be greater than $10d$ and 200mm(d is the minimum diameter among the lapped bars).

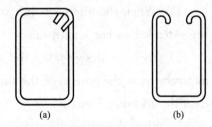

Figure 4-33　Types of double-legged stirrups

(a) closed form; (b) open form

When the diameter of compression bar is greater than 25mm, two stirrups should be set 100mm away from two end sections at the joint of lap.

When the mechanical anchoring measures are adopted, the number of stirrups within the anchorage length shall not be less than 3 and the diameter of the stirrups shall not be less than 0.25 times of the diameter of longitudinal bars. The spacing shall not be more than 5 times of the diameter of the longitudinal reinforcement. When the thickness of the concrete cover for the longitudinal reinforcement is not less than 5 times or equal to the diameter of the reinforcement, no stirrup is needed to be configured.

4.6　Other Structural Requirements of Longitudinal Reinforcement in the Beam and Plate

4.6.1　Longitudinal Reinforced Bars

1. Anchor

(1) The anchorage length l_{as} of the lower ratio longitudinal reinforcement at the support for the simply supported plate and the continuous plate should not be less than $5d$. When temperature and shrinkage stresses in the continuous plate are large, the anchor length into the support should appropriately increase.

(2) There is usually a tensile zone at the upper and compressed at the bottom in the middle support of the continuous beam. The upper longitudinal tensioned bars should extend through the support. The bottom longitudinal reinforcement should be able to withstand the tension when the oblique cracks and bond cracks occur, thus a certain anchor length should be ensured and to be dealt with according to the following circumstances:

1) (When there is no requirement for the longitudinal reinforcement at the bottom) the extended anchor length can be taken as the simply supported case of $V > 0.7 f_t b h_0$;

2) When the design takes full advantage of the tensile strength of the longitudinal reinforcement at the bottom of the support, the extended anchoring length should not be less than the anchorage length l_a;

3) When the design takes full advantage of the compressive strength of the longitudinal reinforcement at the bottom of the support, the extended anchoring length should not be less than $0.7 l_a$.

2. Connection of reinforcement

Connection of reinforcement can be divided into two categories: binding connection, mechanical connection or welding connection.

For the diameter of tensile bar $d > 25$mm and the diameter of compressed bar $d > 28$mm, the lap joints are not allowed in binding type.

(1) Binding lap

For the joint lapped without welding, the lap length l_l is defined as follows:

1) The lap joint of tensile reinforcement

The lap length of tensile reinforcement should be calculated by the following formula on the basis of the percentage of the lap area within the same connected range and not be less than 300mm.

$$l_l = \xi_l l_a \tag{4-27}$$

Where l_l——the lap length of tensile reinforcement;

l_a——the anchoring length of tensile reinforcement;

ξ_l——the correction factor of the lap length for tensile reinforcement according to Table 4-3.

The binding lap joints of adjacent longitudinal reinforcement for one component should be staggered.

The correction factor of the lap length of tensile reinforcement　　　Table 4-3

The area percentage of lap joint for longitudinal reinforcement (%)	≤25	50	100
The correction factor of the lap length ξ_l	1.2	1.4	1.6

The length of connected range is 1.3 times the length of the lap. The lap joints whose midpoint is within the length of the connected range belong to the same connected section, as shown in Figure 4-34.

In any case, the lap length of the binding lap joint for longitudinal tensile reinforcement shouldn't be less than 300mm.

2) The lap joint of compressed reinforcement

The lap length in compression takes 0.7 times the tensile lap length.

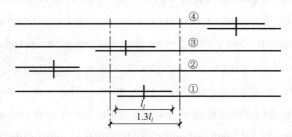

Figure 4-34　The binding lap joint of longitudinal tensile reinforcement within the same connected range

Note: The number of binding lap joints within the same connected range is two in the above figure. When the diameters of reinforcements are the same, the area percentage of lap joints is 50%.

In any case, the lap length of compressed reinforcement shouldn't be less than 200mm.

(2) Mechanical connection or welding

There are many kinds of mechanical connections, among which the cold-rolled straight threaded sleeve connection is more frequently used in our country at present.

The welded joints of adjacent longitudinal reinforcement should be staggered.

The length of the connected range for welded joints is $35d$ (d is the largest diameter of the longitudinal reinforcing bar) and is not less than 500mm. The welded joints whose mid-points within the length of the connecting range belong to the same connected range.

For the joint of longitudinal tensile reinforcement, the area percentage of welded joints in the same connected range should not be greater than 50%, and the area percentage for the longitudinal compressed reinforcement is unrestricted.

4.6.2 Handling Reinforcement and Longitudinal Structural Reinforcement

1. Handling reinforcement

When the beam's span is less than 4m, the diameter of handling reinforcement in the beam should not be less than 8mm; when the beam's span is $4 \sim 6$m, that should not be less than 10mm; when the beam's span is greater than 6m, that should not be less than 12mm.

2. Longitudinal structural reinforcement

For the web height of the beam $h_w \geqslant 450$mm, longitudinal structural reinforcement should be arranged along the height at both sides of the beam. The cross-sectional area of the longitudinal structural reinforcement for each side (not including the reinforcement at the upper and bottom of beam and the handling reinforcement) should not be less than 0.1% of web's cross-sectional area bh_w, whose spacing should not be greater than 200mm. In addition, the web's height h_w is determined according to the Eq. (4-17). The longitudinal structural reinforcement is configured to suppress the action of free load in the web's height of the beam and the development of vertical cracks caused by the shrinkage of concrete.

For reinforced concrete thin-web beam or the reinforced concrete beam with fatigue requirements, longitudinal structural reinforcement whose diameter is $8 \sim 14$mm and spacing is $100 \sim 150$mm should be configured along two sides in the web at the lower part in half beam's height, and should be configured densely at the lower part and sparsely at the upper part. At the upper part in half beam's height for the web, longitudinal structural reinforcement is set according to the provisions of the ordinary beam.

Questions

4-1 What types of cracks will appear on the simply supported beam shown in Figure

4-35? Sketch the crack patterns.

4-2   The formula of shear capacity along an inclined section takes the form that the shear capacity equals the sum of concrete resistance and reinforcement resistance. Does this mean that these two terms are not correlated?

4-3   Why do inclined cracks appear in reinforced concrete beams under loading?

4-4   Sketch the potential inclined cracks and the directions of propagation in the cantilever beams in Figure 4-36.

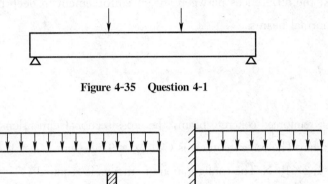

Figure 4-35   Question 4-1

Figure 4-36   Question 4-4

4-5   Why does the flexural failure happen along inclined section? How can this failure mode be prevented?

4-6   What are the main failure patterns along inclined section of simply supported beams with or without web reinforcement? When do they occur? And what are their failure characteristics?

4-7   When should the shear-span ratio be used in the shear capacity design of beams?

4-8   Why should the maximum spacing of stirrups be specified?

4-9   What is the difference between generalized shear-span ratio and computed shear-span ratio? Why can the computed shear-span ratio be used in calculation of shear capacity?

4-10   What are the main factors that will significantly influence the capacities along inclined sections in beams with web reinforcement?

4-11   What is the moment capacity diagram? What is the relationship between the moment capacity diagram and the design moment diagram?

4-12   What is the shear resistance mechanism in a simply supported beam with web reinforcement after the occurrence of inclined cracks?

4-13   Can the shear capacity absolutely be enhanced if more stirrups are provided? Why?

4-14 How would the inclined compression failure be prevented in design?

4-15 What is the difference in the stress state of a simply supported beam without web reinforcement before and after the occurrence of inclined cracks?

4-16 How the influence of the axial forces on the shear resistance of a reinforced concrete column be defined?

4-17 Why is it not correct to simply combine the shear resistance in two perpendicular directions in the calculation of a column under bidirectional shear?

4-18 List the differences between shear reinforcement in deep beams and shear reinforcement in normal beams.

Exercises

4-1 For a reinforced concrete beam, the cross-sectional dimensions are $b \times h = 200\text{mm} \times 500\text{mm}$, $a_s = 35\text{mm}$. The concrete grade is C20 ($f_c = 9.6\text{N/mm}^2$, $f_t = 1.1\text{N/mm}^2$). The applied shear force $V = 1.2 \times 10^5 \text{N}$. The configuration of stirrups is required ($f_{yv} = 270\text{N/mm}^2$).

4-2 All other conditions are the same as those in exercise 4-1 except the shear forces are $V = 6.2 \times 10^4 \text{N}$ and $V = 2.8 \times 10^5 \text{N}$, respectively. Recalculate the stirrups for each force.

4-3 A reinforced concrete beam subjected to uniformly distributed load $q = 40\text{kN/m}$ (self-weight included) is shown in Figure 4-37. C20 concrete ($f_c = 9.6\text{N/mm}^2$, $f_t = 1.1\text{N/mm}^2$) is used. The cross-sectional dimensions are $b \times h = 200\text{mm} \times 400\text{mm}$. The arrangements of the stirrups at cross section including A_{right}, B_{left} and B_{right} ($f_{yv} = 270\text{N/mm}^2$) are required.

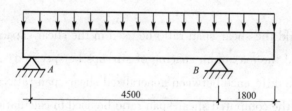

Figure 4-37 Exercise 4-3 (unit: mm)

4-4 For a simply supported beam as shown in Figure 4-38, a uniformly distributed load $q = 70\text{kN/m}$ (self-weight included) is applied. C20 concrete ($f_c = 9.6\text{N/mm}^2$, $f_t = 1.1\text{N/mm}^2$) is used. The strengths for longitudinal reinforcement and stirrups are $f_y = 300\text{N/mm}^2$ and $f_{yv} = 270\text{N/mm}^2$, respectively. Calculate the following: (1) the required stirrups if bent-up bars are not provided; (2) the required stirrups if existing longitudinal bars are utilized as bent-up bars; (3) the configuration of bent-up bars if stirrups of $\Phi 8@200$ are provided.

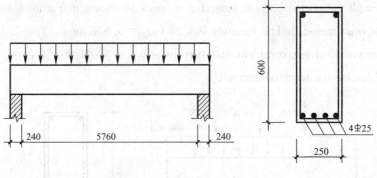

Figure 4-38   Exercise 4-4 (unit: mm)

4-5   A simply supported beam of rectangular cross section is shown in Figure 4-39. The cross-sectional dimensions are $b \times h = 200\text{mm} \times 400\text{mm}$. Concrete of C20 ($f_c = 9.6\text{N/mm}^2$, $f_t = 1.1\text{N/mm}^2$) is used. The strengths for longitudinal reinforcement and stirrups are $f_y = 300\text{N/mm}^2$ and $f_{yv} = 270\text{N/mm}^2$, respectively. If the self-weight of the beam is ignored, try to calculate the following: (1) the required longitudinal tensile bars; (2) the required stirrups (without bent-up bars); (3) the required stirrups if longitudinal bars are utilized as bent-up bars.

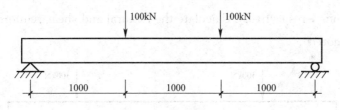

Figure 4-39   Exeruse 4-5 (unit: mm)

4-6   For a simply supported reinforced concrete beam as shown in Figure 4-40, a uniformly distributed load $q = 70\text{kN/m}$ (self-weight included) is applied. Concrete of C20 ($f_c = 9.6\text{N/mm}^2$, $f_t = 1.1\text{N/mm}^2$) is used. The strengths for longitudinal reinforcement and stirrups are $f_y = 300\text{N/mm}^2$ and $f_{yv} = 270\text{N/mm}^2$, respectively. Is the beam safe under such conditions?

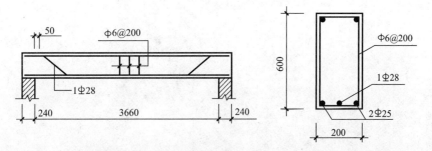

Figure 4-40   Exercise 4-6 (unit: mm)

4-7　A simply supported beam is subjected to loads as shown in Figure 4-41, in which the self-weight has been included. The concrete is C20 ($f_c = 9.6\text{N/mm}^2$, $f_t = 1.1\text{N/mm}^2$). The strengths for longitudinal reinforcement and stirrups are $f_y = 300\text{N/mm}^2$ and $f_{yv} = 270\text{N/mm}^2$, respectively. Calculate the required stirrups.

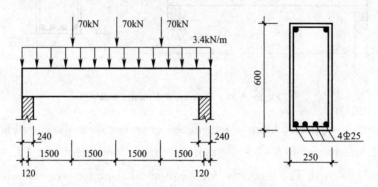

Figure 4-41　Exercise 4-7 (unit: mm)

4-8　A two-span continuous beam with the cross-sectional dimensions of $b \times h = 250\text{mm} \times 500\text{mm}$ is shown in Figure 4-42. The concrete is C20 ($f_c = 9.6\text{N/mm}^2$, $f_t = 1.1\text{N/mm}^2$). The strengths for longitudinal reinforcement and stirrups are $f_y = 300\text{N/mm}^2$ and $f_{yv} = 270\text{N/mm}^2$, respectively. Calculate the flexural and shear reinforcement (the self-weight can be ignored).

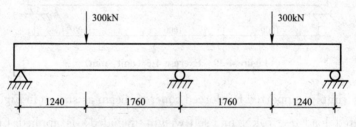

Figure 4-42　Exercise 4-8 (unit: mm)

Chapter 5
Sectional Load-Carrying Capacity for Compression Members

5.1 Introduction

The members that are mainly subjected to axial compression are called the compression members. Typical examples in reinforced concrete structures include columns of single-story industrial building, arches, upper chords in trusses, frame column of multistory structures and high-rise buildings, shear wall, chimney wall, bridge piers, piles, etc. The compressiom members can be divided into the axial compression members, the uniaxial eccentric compression members and the biaxial eccentric compression members. Members subjected to axial compression at the geometric centers of cross sections are called axial compression members. The members subjected to the axial compression force with the action point deviating from one of the main axes of member's cross section are called uniaxial eccentric compression members, or else members with action points deviating from the two main axes are called biaxial eccentric compression members.

5.2 Details of Compression Member

5.2.1 Section Type and Dimensions

To facilitate the templates work, square or rectangular sections are commonly used in axially loaded columns. Circular and regular polygonal sections can be used under special requirements. For eccentric compression members, the rectangular sections are commonly used. However, in order to reduce the weight of columns, I-sections are common types for bigger columns, especially in prefabricated columns. Circular sections are chosen for chimneys, water tower pillars and columns made by centrifugal principle.

For columns with square and rectangular sections, the section dimension should not be smaller than 250mm×250mm. To ensure the stability of columns, l_0/b should not be greater than 30 and l_0/h should not be greater than 25. Where l_0 is the calculated length of the column, b and h are the shorter side and longer side of a rectangular section, respectively. For convenience in setting up the formwork during construction, section dimensions of a column should be integers. If it is smaller than 800mm, it should be taken as the multiples of 50mm. Otherwise, it should be taken as the multiples of 100mm. For I-section and T-section columns, the flange thickness and web thickness of cross sections should not be smaller than 120mm and 100mm, respectively.

5.2.2 Materials

The concrete strength has an evident effect on the load-carrying capacity of the compression members. In order to reduce the section dimensions and save the amount of steels, the concrete with the grade of C30, C35 and C40 are generally recommended. For bottom columns of high-rise buildings, much higher grade concrete may be used.

The longitudinal reinforcement with the grade of HRB400, RRB400 and HRB500 are generally adopted in concrete structures. HRB400, HRB335 and HPB300 are generally used as stirrups.

5.2.3 Longitudinal Steel Bars

The diameter of longitudinal steel reinforcement in column should not be less than 12mm, which generally ranges from 12mm to 32mm. The reinforcement ratio of all longitudinal steel bars should not be larger than 5%, and longitudinal reinforcement ratio on one side of the section should not be less than 0.2%.

For axial compression members, longitudinal reinforcements should be evenly distributed along all sides of a section. The number of bars should not be less than 4 as shown in Figure 5-1(a). The amount of longitudinal steel bars applied in circular column may not be less than 8 and should not be less than 6, which should be uniformly arranged along the periphery of the column section. To avoid transverse bending of the bars, steel bars with larger diameters are recommended. For eccentric compression members, longitudinal reinforcements should be arranged on both sides of the bending moment direction. If the cross-section height h is equal to or greater than 600mm, additional longitudinal bars with diameters of 10~16mm should be set on the lateral sides as shown in Figure 5-1(b).

The clear spacing of longitudinal steel bars in columns should not be less than 50mm and may not be larger than 300mm. In eccentric compression columns, the center-to-center distance of longitudinal steel bars on the side perpendicular to the acting plane of bending moment as well as the center-to-center distance of longitudinal steel reinforcements on each side in the axial compression columns may not be larger than 300mm.

5.2.4 Stirrups

To fix the longitudinal reinforcement bars and prevent the bars from buckling, stirrups in the columns should be made as closed forms. For a bound and welded skeleton, the spacing of the stirrups along the length of the member should not be greater than $15d$ and $20d$, where d is the minimum diameter of the longitudinal steel bars. Meanwhile, the spacing should not be greater than the shorter side of the section or 400mm.

The diameter of the stirrup should not be less than 1/4 of the maximum diameter of the longitudinal steel bars, and should not be less than 6mm.

When the longitudinal reinforcement ratio is greater than 3%, the diameter of stirrups should not be less than 8mm. The spacing of stirrups should not be greater than $10d$ or 200mm. The end of stirrups should be properly anchored. The stirrups may be welded into close hoops, or have both ends bent with a 135° angle hook and embedded in the core of member with a length not less than 10 times of the diameter of the stirrup.

When the short side of the cross section is greater than 400mm and the number of longitudinal steel bars is over 3, or one side is not greater than 400mm but the number of longitudinal bars is over 4, additional lateral stirrups will be needed besides the closed stirrups as shown in Figure 5-1(b).

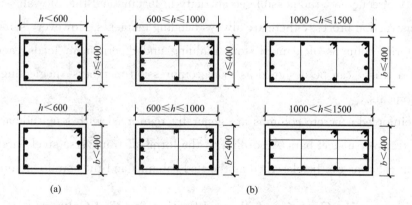

Figure 5-1 Stirrups in rectangular section
(a)$h<600$; (b)$h\geqslant 600$

It is appropriate to set longitudinal reinforcements at every corner of the stirrups.

In the range of longitudinal reinforcement lap length, the diameter of the stirrup should not be less than 0.25 times the diameter of the lap reinforcement; the stirrup spacing should not be greater than $5d$ or 100mm, where d is the smallest diameter of the lap reinforcement. When the overlapped steel bars are more than 25mm in diameter, two stirrups should be set in the range of 100mm outside the two ends of the lap joint.

For a column with a complex cross-section, stirrups with inner corner as shown in Figure 5-2 should not be used.

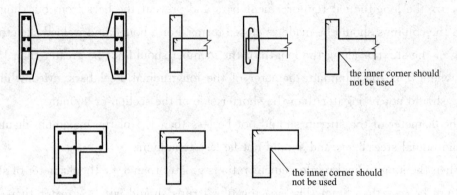

Figure 5-2 Stirrups in I-shaped, L-shaped sections

5.3 Compression Capacity of Normal Section for Axial Compression Members

In engineering practice, due to the existences of initial bending in members, the non-homogeneity of concrete, the possible eccentricity in loading and the allowable errors during the construction process, there are always bending moments. However, some members such as interior columns of a multi-story building under permanent loads, and the web member of a truss, can be designed as axial compression members, neglecting the small bending moments.

The reinforced concrete columns are generally reinforced with longitudinal reinforcements and transverse steel bars. According to the forms of transverse steel bars, the axial compression columns can be classified as the tied columns and the spiral columns.

5.3.1 Compression Capacity of Normal Section for Tied Columns

The most common axial compressive columns are tied columns as shown in Figure 5-3. The longitudinal reinforcements are ranged along the periphery of the cross section with the following capacities: providing resistance to axial compression and reducing the dimensions of a section, resisting some bending moments caused by accidental eccentricity, reducing the effects of the creep and shrinkage of concrete under sustained compressive stresses on members and increasing the ductility of the member.

Transverse steel bars are as important as longitudinal reinforcements with the following effects on a member: fixing the positions of longitudinal steel bars, preventing the buckling of reinforcements and confining the lareral expansion of the compressed concrete.

1. Force analysis and destruction form

For a tied short column under an axial load, test results have shown that the strain distribution across the section is essentially uniform. When the load is small, the concrete and the steel are in the elastic stage. The increase of the compressive deformation of the column is proportional to the increase of the load. The increase of the compressive stress of the longitudinal reinforcement and the concrete is also proportional to the increase of the load. With load increment, the increase of deformation is greater than that of load as shown in Figure 5-4. When

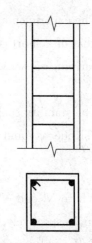

Figure 5-3 The column with longitudinal reinforcements and stirrups

the strain of concrete reaches the value of 0.0025~0.0035, the longitudinal cracks appear on the surface of the column and the concrete cover begins to split. Then longitudinal reinforcements are buckled and the concrete is crushed. Figure 5-5 illustrates the typical failure mode of a column.

For slender columns, experimental results illustrate that the effects of the initial eccentricity caused by various contingencies can not be neglected. When the load is increased, the eccentricity causes lateral deformation of the column. The deflection will also increase the eccentricity of the load. As the load increases, the effects of the additional bending moment and lateral deflection will become more significant until the concrete is crushed and the longitudinal steel bars are bowed and convex as shown in Figure 5-6.

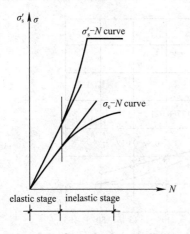  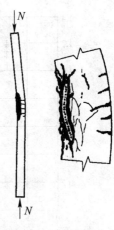

Figure 5-4 Stress-load curve diagram   Figure 5-5 Failure of a short-column   Figure 5-6 Failure mode of a slender column

The experimental results also show that the ultimate load of a slender column is lower than that of a short column. The larger the slenderness ratio of column is, the smaller the load-carrying capacity will be. If a column is extremely slender, an instability failure may occur.

The stability coefficient φ is used to represent the reduction of the load-carrying capacity of slender column in *Code for Design of Concrete Structures*.

$$\varphi = \frac{N_u^l}{N_u^s} \tag{5-1}$$

Where N_u^l and N_u^s are the load-carrying capacities of the slender column and the short column, respectively.

Test results have shown that the coefficient φ is mainly dependent on the slenderness ratio of a compressive member as shown in Figure 5-7. The slenderness ratio is defined as the ratio of the effective length l_0 of a column to the minimum radius of gyration i of the section, and i is replaced by the length of the shorter side b in the rectangular section.

According to the experimental data and statistics, the following empirical formulae are available as:

When $l_0/b = 8 \sim 34$

$$\varphi = 1.117 - 0.012 l_0/b \tag{5-2}$$

When $l_0/b = 35 \sim 50$

$$\varphi = 0.87 - 0.012 l_0/b \tag{5-3}$$

The values of coefficient φ are recommended in *Code for Design of Concrete Structures* GB 50010—2010 as shown in Table 5-1. In practical design, the effective lengths of reinforced concrete columns are specified in section 7.3.11 of GB 50010—2010.

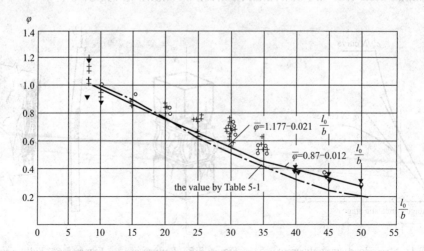

Figure 5-7 Test results and standard values of the stability coefficient

Stability Coefficient of Reinforced Concrete Members Table 5-1

l_0/b	l_0/d	l_0/i	φ	l_0/b	l_0/d	l_0/i	φ
≤8	≤7	≤28	≤1.00	30	26	104	0.52
10	8.5	35	0.98	32	28	111	0.48
12	10.5	42	0.95	34	29.5	118	0.44
14	12	48	0.92	36	31	125	0.40
16	14	55	0.87	38	33	132	0.36
18	15.5	62	0.81	40	34.5	139	0.32
20	17	69	0.75	42	36.5	146	0.29
22	19	76	0.70	44	38	153	0.26
24	21	83	0.65	46	40	160	0.23
26	22.5	90	0.60	48	41.5	167	0.21
28	24	97	0.56	50	43	174	0.19

Note: l_0 is the effective length of a column; b is the short side of a rectangular section; d is the diameter of a circular section; i is the minimum radius of gyration of the section.

2. Formula for compression capacity of normal section

For axial compression members reinforced by moderate strength steel bars, when the compressive stress of concrete reaches its maximum and the steel yields (as shown in Figure 5-8), the members are assumed to reach their ultimate compressive capacities. The formula to calculate the ultimate capacity is:

$$N_u = 0.9\varphi(f_c A + f'_y A'_s) \quad (5-4)$$

Where N_u——ultimate compressive capacity of the column;

0.9——reliability adjustment factor;

φ——stability coefficient;

f_c——axial compressive strength of concrete;

A——sectional area of the column;

A'_s——sectional area of all longitudinal bars.

When the ratio of longitudinal steel bars is more than 3%, the net sectional area of concrete A_n should replace A in Eq. (5-4), where $A_n = A - A'_s$.

Figure 5-9 shows the redistribution of stresses for the concrete and steel bars under long-term load. The rate of reduction for stress in concrete decreases gradually and the rate of increase for stresses in steel bars also decreases gradually. The variation degree for stress in concrete is less than that in steel.

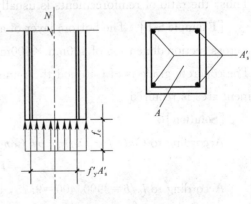

Figure 5-8 Calculation diagram for cross section compression capacity of the tied columnns

The smaller the ratio of reinforcements is, the larger the variation in steel stress is.

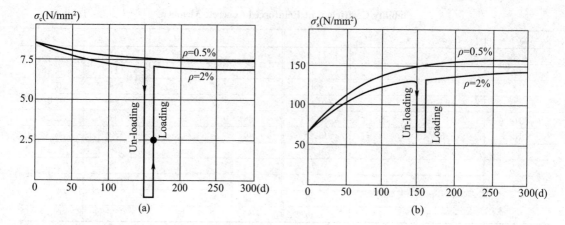

Figure 5-9 Redistribution of stresses for concrete and steel bars uder long-term load

(a) The concrete diagram; (b) The steel bars diagram

For a column under sustained loads, tests have shown that creep and shrinkage of concrete cause the load to transfer from the concrete to the reinforcements. Figure 5-9 also shows how the stresses in concrete and steels vary over time.

For a column with a high percentage of reinforcements under a huge initial load, if the load is entirely removed, the strain will disappear immediately, but the creep deformation will not. Self-equilibrium stresses will be formed in the section by the creep deformation of concrete. This is because when concrete shrinks due to creep, embedded steel bars are compressed. When the load is removed suddenly, tensile stress in surrounding concrete will be generated. If the same load is applied to the column again, the stresses in the concrete and the steel will return to the previous stress stage. If reinforcements exceed a certain amount, the tensile stress of the concrete may reach the strength f_t and the column may be cracked. Thus, the ratio of reinforcements is usually no more than 5% in design.

[Example 5-1] The bottom layer of a four-story cast-in-situ frame with four-span has a cross-section dimension of 400mm×400mm, under an axial load $N=3600$kN, $H=3.9$m. The concrete grade is of C40 and the reinforcement is of HRB400. Longitudinal reinforcement area is required.

[Solution]

According to *Code for Design of Concrete Structures* GB 50010

$$l_0 = H = 3.9 \text{m}$$

According to $l_0/b = 3900/400 = 9.75$, look up in Table 5-1, $\varphi=0.983$

According to Eq. (5-4), A_s' can be obtained:

$$A_s' = \frac{1}{f_y'}\left(\frac{N}{0.9\varphi} - f_c A\right) = \frac{1}{360}\left(\frac{3600 \times 10^3}{0.9 \times 0.983} - 19.1 \times 400 \times 400\right) = 2814 \text{mm}^2$$

7 ⌽ 22 is adopted, $A_s' = 2661 \text{mm}^2$

$\rho' = \dfrac{A'_s}{A_s} = \dfrac{1256}{400 \times 400} = 1.7\% < 3\%$, therefore, there is no need to take the net sectional area A_n, $\rho'_{min} = 0.6\%$, $\rho > \rho'_{min}$, it is under safety.

The reinforcement rate for cross section on each side
$$\rho' = \dfrac{0.5 \times 2661}{400 \times 400} = 0.8\% > 0.2\%$$

Hence the column is under safety (the total longitudinal reinforcement requirement is $\rho'_{min} = 0.6\%$; one side of the minimum longitudinal reinforcement ratio is $\rho' = 0.2\%$) by taking 7 Φ 22 of $A'_s = 2661\text{mm}^2$.

[**Example 5-2**] The height for a ground floor interior column in reinforced concrete frame without sidesway from the foundation level to the top of the slab is 2.8m, the column has a square section of $b = h = 250$mm. The concrete grade is of C40, the reinforcements are taken 4 Φ 22 bars of HRB400 arranged symmetrically in the section. If the axial load is 1500kN, the calculation for the safety of this member is required.

[**Solution**]

According to $l_0/b = 2800/250 = 11.2$, look up in Table 5-1, $\varphi = 0.962$

According to Eq. (5-4)
$$N_u = 0.9\varphi(f_c A + f'_y A'_s) = 0.9 \times 0.962 \times (19.1 \times 250 \times 250 + 360 \times 1520)$$
$$= 1507 \times 10^3 \text{N} > N = 1500 \times 10^3 \text{N}$$

So the cross section is safe.

5.3.2 Compression Capacity of Normal Section for Spiral Columns

When the column is subjected to a large axial compression force and the column section size is limited by the requirements of the building and the common use, it is not sufficient to bear the load by reinforcing ordinary stirrups. Therefore the use of spiral stirrups or welding rings can be considered to improve the bearing capacity. The cross-sectional shape of the column is generally circular or polygonal and Figure 5-10 shows the configuration of the spiral stirrups and the welded rings.

Spiral stirrups and welding ring ribs with high stirrups are not likely to "collapse" like ordinary stirrups, constraining the lateral deformation of the core concrete in the longitudinal compression, which

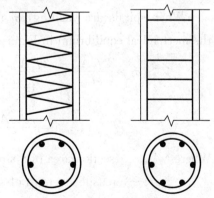

Figure 5-10 Spiral stirrups and welding rings

increase the compressive strength of the concrete and deformation capacity. This constrained concrete is called "confined concrete". At the same time, the spiral stirrups or welding ring ribs bear tensile stress. When the external force gradually increases, the stirrup's stress reaches the tensile yield strength, leading to lose the capacity to effectively restrain the lateral deformation of concrete and improve the concrete compressive strength. Then the component approaches its damage. It can be seen that the use of spiral stirrups or welding rings in the transverse direction of the columns can also be functioned as a part of the longitudinal reinforcement so as to improve the loading capacity and the deformability, which is called "indirect reinforcement". The protective layer of concrete outside the spiral stirrups or the welded ring bars will crack or collapse when the spiral stirrups or welded ring bars are subjected to high tensile stress. Therefore, this part of the concrete is not considered in the calculation.

As we all know, for shear, torsion and punching resistance design, the tensile strength design value of stirrups is limited to no more than 500MPa. However, when acting as the indirect reinforcement of concrete (such as continuous spiral hoops or closed welding hoops), its strength can be fully developed, such as the use of 500MPa grade steel or higher strength steel, which shows a certain economic benefits.

According to the above analysis, it can be seen that the core cross-section concrete surrounded by helical stirrups or welded ring ribs is in the three-way compression state so that the axial compressive strength is higher than the uniaxial compressive strength, the calculation for the strength improvement can be obtained:

$$f = f_c + \beta \sigma_r \tag{5-5}$$

Where f——axial compressive strength of confined concrete;

σ_r——the radial compressive stress value in the core concrete of the column.

When spirals are yielding, σ_r reaches its maximum. By the mechanical equilibrium in Figure 5-11, σ_r can be obtained

$$\sigma_r = \frac{2f_y A_{ssl}}{s d_{cor}} = \frac{2f_y A_{ssl} d_{cor} \pi}{4\pi \frac{d_{cor}^2}{4} s} = \frac{f_y A_{ss0}}{s d_{cor}} \tag{5-6}$$

$$A_{ss0} = \frac{\pi d_{cor} A_{ssl}}{s} \tag{5-7}$$

Where A_{ssl}——section area of a single spiral;

f_y——tensile strength of spirals;

s——pitch of spirals;

d_{cor}——diameter of the core concrete, generally $d_{cor} = d_c - 2c$, where d_c stands for the diameter of a

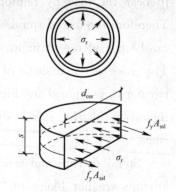

Figure 5-11 Radial pressure diagram of concrete

column and c is the thickness of concrete cover;

A_{ss0}——equivalent area of lateral reinforcement;

A_{cor}——sectional area of core concrete.

From equilibrium in longitudinal direction, the ultimate capacity of a spiral column can be calculated as:

$$N_u = (f_c + \beta\sigma_r)A_{cor} + f'_y A'_s \tag{5-8}$$

Taking $2\alpha = \beta/2$ into the above equation, and considering the adjustment factor 0.9, the *Code for Design of Concrete Structures* stipulates that:

$$N_u = 0.9(f_c A_{cor} + 2\alpha f_y A_{ss0} + f'_y A'_s) \tag{5-9}$$

When the cube strength of concrete is not larger than 50N/mm^2, $\alpha = 1.0$. When the cube strength of concrete is 80 N/mm^2, $\alpha = 0.85$. α can be determined by a linear interpolation for concrete with cube strength between 50MPa and 80MPa.

In order to ensure that the concrete protective layer on the outside of the indirect reinforcement is sufficiently safe to resist the shedding, the bearing capacity of the member calculated by the Eq. (5-9) should not be 50% larger than that of the Eq. (5-4).

The following items should be noted when using Eq. (5-9):

(1) When $l_0/d > 12$, due to the large slenderness, the beneficial effect of lateral reinforcement will be neglected;

(2) The calculated ultimate load of a spiral column section is less than that of a tied column section;

(3) When the equivalent area of lateral reinforcement A_{ss0} is smaller than 25% of the total area of longitudinal steel bars, lateral reinforcement is assumed to be insufficient to provide enough confinement to core concrete.

The code stipulates that the spacing of the spirals along the length of a column should not be less than 40mm and greater than 80mm or $d_{cor}/5$. The diameter of a spiral bar should not be less than 6mm or 1/4 of the maximum diameter of longitudinal bars.

[**Example 5-3**] A cast-in-place reinforced concrete column of a hotel in the main hall under the class 1 of environment is subjected the axial compression with the design value $N = 6000\text{kN}$. The height from the base to the top of the surface is $H = 5.2\text{m}$. A cylindrical section is designed with the diameter $d = 470\text{mm}$. The concrete grade is of C40, the longitudinal reinforcement is of HRB400 and the stirrups are of HPB300.

The spiral bars and the reinforcements for the member are required.

[**Solution**]

Firstly regular longitudinal reinforcement and stirrups are designed.

(1) Calculate the length of the bottom column of the cast-in-situ frame $l_0 = H = 5.2\text{m}$.

(2) The stability factor φ
$$l_0/d = 5200/470 = 11.06$$
Look up in Table 5-1, $\varphi=0.938$

(3) Calculate longitudinal reinforcement area A_s'

The cross-sectional area is
$$A = \pi d^2/4 = 3.14 \times 470^2/4 = 17.34 \times 10^4 \text{mm}^2$$
From Eq. (5-4)
$$A_s = \frac{1}{f_y}\left(\frac{N}{0.9\varphi} - f_c A\right) = \frac{1}{360}\left(\frac{6000 \times 10^3}{0.9 \times 0.938} - 19.1 \times 17.34 \times 10^4\right) = 10543 \text{ mm}^2$$

(4) Calculate the reinforcement ratio
$$\rho' = A_s'/A = 10543/(17.34 \times 10^3) = 6.08\% > 5\%$$

Reinforcement ratio is too high. Considering $l_0/d<12$, spiral stirrups can be used. The following procedures are based on spiral stirrups design.

(5) Assume that the longitudinal reinforcement ratio $\rho'=0.045$, $A_s=\rho'A=7803\text{mm}^2$, 16 Φ 25 with $A_s'=7854\text{mm}^2$ is taken. The concrete cover is in thickness of 20mm and the stirrup diameter is estimated to be 10mm.
$$d_{cor} = d - 30 \times 2 = 470 - 60 = 410 \text{mm}$$
$$A_{cor} = \pi d_{cor}^2/4 = 3.14 \times 410^2/4 = 13.2 \times 10^4 \text{ mm}^2$$

$\alpha=1.0$ is for concrete grade lower than C50; according to Eq. (5-9), spiral stirrups equivalent area A_{ss0} is:
$$A_{ss0} = \frac{N/0.9 - (f_c A_{cor} + f_y' A_s')}{2f_y}$$
$$= \frac{6000 \times 10^3/0.9 - (19.1 \times 13.20 \times 10^4 + 360 \times 7854)}{2 \times 270} = 2441 \text{mm}^3$$

$A_{ss0} > 0.25 A_s' = 0.25 \times 7854 = 1964 \text{mm}^2$, the structural requirement is met.

The stirrups' diameter is $d=10$mm, then the single-limb spiral stirrups' area is $A_{ss1}=78.5\text{mm}^2$. The stirrup spacing s can be obtained according to Eq. (5-7):
$$s = \pi d_{cor} A_{ss1}/A_{ss0} = 3.14 \times 410 \times 78.5/2441 = 41.4 \text{mm}^2$$

Take $s=40$mm, the requirements that the spacing is not less than 40mm and not more than 80mm or $0.2d_{cor}$ are satisfied.

(6) According to the helical stirrups with $d=10$mm, $s=40$mm, Eqs. (5-7) and (5-9) are combined to get the design value N_u as follows:
$$A_{ss0} = \frac{\pi d_{cor} A_{ss1}}{s} = \frac{3.14 \times 410 \times 78.5}{40} = 2527 \text{mm}^2$$
$$N_u = 0.9(f_c A_{cor} + 2\alpha f_y A_{ss0} + f_y' A_s')$$
$$= 0.9(19.1 \times 13.20 \times 10^2 + 2 \times 1 \times 270 \times 2527 + 360 \times 7854)$$
$$= 6041.88 \text{kN}$$

From Eq. (5-4)

$$N_u = 0.9 \cdot \varphi \cdot (f_c \cdot A + f'_y \cdot A'_s)$$
$$= 0.9 \times 0.938 \times [19.1 \times (17.34 \times 10^4 - 7854) + 360 \times 7854]$$
$$= 5056.23 \text{kN}$$

And $1.5 \times 5056.23 = 7584.35 \text{kN} > 6041.88 \text{kN}$, the requirement is met.

5.4 Failure Modes of Normal Section for Eccentric Compression Members

5.4.1 Failure Modes of Eccentric Compressive Short Columns

From experimental results, there are two major failure modes for a member under an eccentric load. The amounts of tension reinforcing bars and relative eccentricity (e_0/h_0) have obvious effects on the modes of failure.

1. Tensile Failure (or large eccentric compression failure)

The form of tensile failure is also known as the large eccentric compression failure, which occurs when the relative eccentricity of the axial compression force N is large and the tensile reinforcement ratio is not high. Under this condition, the stress is tensile on one side and compressive on the other. With the load increase, the transverse cracks are firstly generated in the tension zone. Under the load continuously increasing, the number of cracks are increasing. The main cracks are gradually wide before the damage. The stress of the tensile bar reaches the yield strength and enters into the yield stage. The development of the deformation in tensile zone is greater than the compression deformation. With the neutral axis approaching the compressive zone, the height of the compressive zone quickly shortens till the edge of the concrete reaches its ultimate bearing capacity. After the longitudinal steel bars in compression yield, the component is finally destroyed. The type of damage is a ductile pattern and the stress of the longitudinal reinforcement in the compressive zone can also reach the yield strength. Briefly the characteristics of the tensile failure is that the tensile reinforcement first reaches the yield strength and the edge of the concrete in compression crush ultimately.

This failure pattern is similar to the stiffened beam. When the member is broken, the stress state on the normal section is shown in Figure 5-12(a) and Figure 5-12 (b).

2. Compression failure (or small eccentric compression failure)

Compression failure is also known as small eccentric compression damage. The section damage is from the edge of the compression zone in the following two cases.

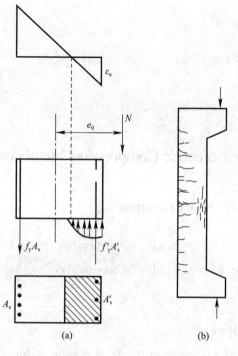

igure 5-12 The tensile stress and the form of the tensile failure
(a) Section stress;
(b) The form of the tensile failure

The first case: when the axial force N relative to the eccentricity is small, the section bears all the compression or most of the compression, as shown in Figure 5-13(a) and (b). In general, the failure of the section occurs initially by the yielding of compressive reinforcement. Then the strain in the extreme compressive fiber of the concrete near the eccentricity reaches its ultimate compressive strain and the concrete is crushed. The reinforcements on the opposite side of the eccentricity may be either in tension or in compression, but the steel bars do not reach their yield strength, as shown in Figure 5-13 (a) and (b). Only when the eccentricity is very small (for the rectangular section $e_0 \leqslant 0.15h_0$) and the axial force N is very large ($N > \alpha_1 f_c bh_0$), the reinforcement far away from the eccentricity may also yield. In addition, if relative eccentricity is very small, the failure of the section may occur initially by the crushing of concrete on the opposite side of the eccentricity, which can be called "reverse damage".

The second case: when the relative eccentricity of the axial force N is large, but with a large number of tensile steel bars not yielding. In this case of failure, the concrete at the edge of the compression zone reaches the ultimate compressive strain. The stress of the compression bar reaches the compressive yield strength but the reinforcements in the other side are tensioned without yielding. The stress state on the cross section is shown in Figure 5-13 (a). There is no obvious sign before damage, the crushing scale in height is long. The higher the strength of concrete is, the more sudden the damage will be, as shown in Figure 5-13 (c).

Briefly, the characteristics of the compressive failure (or small eccentric compression failure) is that the concrete crushes first and the opposite side of the steels do not yield under tension or compression. This belongs to the brittle damage.

In summary, the "tension failure" and "compressive failure" are both based on the destruction of the material. Obviously a balanced failure between the compression failure and the tension failure exists, which is characterized by the simultaneous occurrence of the concrete crushing and the longitudinal steel bars' yielding; the reasons for the two damages

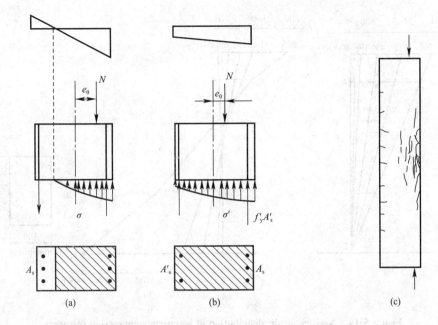

Figure 5-13 The stress distribution and compression failure
(a), (b) Section stress; (c) The form of the compressive failure

are different, the tension failure is caused by the tensile steel yielding while the compressive failure is due to the crushing of the concrete. The balanced damage also belongs to the form of tensile failure.

The test results have shown that from the beginning of loading to the final damage, the strains measured along the height of the section can fit the plane section assumption well. Figure 5-14 presents the two eccentric compression specimens, the average strains along the cross-section height are approximately linearly distributed.

5.4.2 Failure Modes of Eccentric Compressive Slender Columns

The test results show that the reinforced concrete column will produce longitudinal bending under eccentric compression. However, to the small slenderness ratio column, known as "short column", the additional eccentricity can be negligible, because the bending deformation is very small. Comparatively, the additional eccentricity of a slender column must be considered in design, just because the lateral deflection is rather large. Figure 5-15 shows the load-lateral deformation (N-f) test curve for a long column.

Under the influence of longitudinal bending, the long column under the eccentric compression may be damaged in two types. When the slenderness ratio is large, the damage of the member is not caused by the material destruction, but by the longitudinal bending till to the member loss of balance, which is called "buckling failure". When the column length

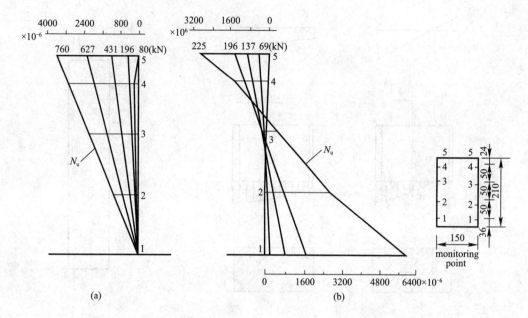

Figure 5-14 Average strain distribution of eccentric compression members

(a) Compressive failure $\frac{e_0}{h_0}=0.24$; (b) Tensile failure $\frac{e_0}{h_0}=0.68$

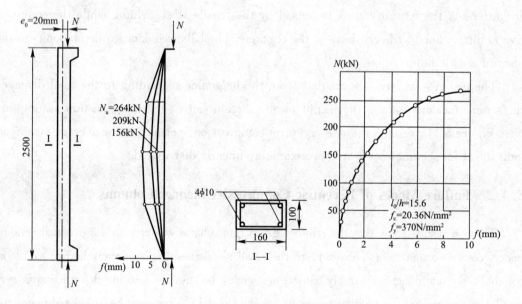

Figure 5-15 $N\text{-}f$ curve of a slender column

is within a certain range, the eccentricity increases from e_i to e_i+f, leading to the bearing capacity of the column smaller than that in a short column with the same section. This failure type belongs to the "material damage".

Figure 5-16 compares the behaviors of three reinforced concrete columns of the same section and same longitudinal reinforcements but with different slenderness under increasing loads.

The curve $ABCD$ in Figure 5-16 shows the correlations between the bearing capacity M and N at failure. The straight line OB represents the relationship between N and M for the short column with small slenderness ratio from loading to the point B of failure.

Since the longitudinal bending of the short column is very small, it can be assumed that the eccentricity is constant from beginning to the end, that is to say, M/N is constant with a straight line, as is called "material damage".

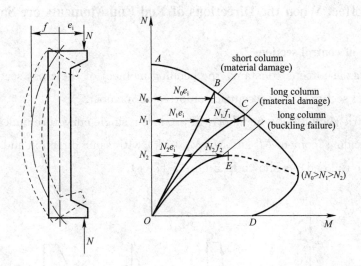

Figure 5-16 N-M curve between columns with different slenderness ratios

Curve OC is the relationship between N and M for the long column from loading to the point C of failure. In the long column, the eccentricity increases with the increasing of the longitudinal force, that is to say, the M/N is varying as a curve but it is also within the "material damage". When the slenderness ratio is larger, the column can be destroyed by the increase of the non-convergent bending moment M, much earlier before the $ABCD$ of the material failure. This is "buckling failure", as shown in curve OE. The bearing capacity at point E is the ultimate load with no steel bars reaching the yield strength. Meanwhile, the concrete strain does not reach the ultimate compressive strain as well. Figure 5-16 has shown the three columns' eccentricities e_i are the same, but their abilities to withstand the longitudinal force N are different, namely, $N_0 > N_1 > N_2$, indicating that the increasing of slenderness ratio will reduce compression bearing capacity. When the slenderness ratio is large, the bending of the eccentric compression member will cause the loss of balance. The reason for this phenomenon is that when the slenderness is large, the longitudinal bending of the eccentric compression member causes a non-negligible additional moment or second-order effect.

5.5 P-δ Effect of Eccentric Compression Members

The additional bending moment and additional curvature caused by the axial force in the member with lateral displacements and deflections are called the second-order effect. Among them the second-order effect produced by the deflection is called the P-δ effect.

5.5.1 P-δ Effect When the Directions of Rod End Moments are Same

1. The transfer of control sections

When the axial force is constant, the bending moment of the cross section will determine the control section in the eccentric compression member.

The eccentric compression member will produce a single curvature under the combined action of the bending moment M_1 and M_2 ($M_2 > M_1$) with same direction and the axial force P applied at the rod end, as shown in Figure 5-17 (a).

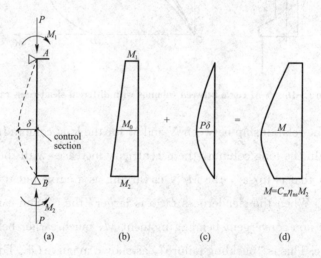

Figure 5-17 The second-order effect (P-δ effect)

When the second-order effect is not taken into account, the bending moment of the member is shown in Figure 5-17 (b), and the largest bending moment is M_2 of the section B, which is determinative to the bearing capacity.

After considering the second-order effect, the axial force P has an additional bending moment $P\delta$ for any section of the middle part of the member, and the bending moment is obtained by integrating the first order moment M_0 with the additional moment:

$$M = M_0 + P\delta \tag{5-10}$$

Where δ——the deflection value of any section.

Figure 5-17(c) and (d) show an additional bending moment diagram and synthetic bending moment diagram, respectively. It can be seen that there exists a cross section with the largest bending moment M in the middle of the member. If the additional bending moment $P\delta$ is relatively large and M_1 is close to M_2, there may be the situation of $M>M_2$. At this time, the control section of the eccentric compression member is transferred from the original rod end section to the section with the largest bending moment in the middle of the member. For example, when $M_1=M_2$, it can be calculated that the control section is at the midpoint of the length of the member.

It can be concluded that the control section is transferred to the middle of the length of the rod, due to the second-order effect of P-δ.

2. Conditions to consider the *P-δ* effect

In order to reduce the calculation errors, the *Code for Design of Concrete Structures* provides specifications to consider the P-δ second-order effect under the following circumstances:

$$M_1/M_2 > 0.9 \tag{5-11a}$$

$$\text{Axial compression ratio } N/f_c A > 0.9 \quad \text{or} \tag{5-11b}$$

$$\frac{l_c}{i} > 34 - 12(M_1/M_2) \tag{5-11c}$$

Where M_1, M_2——M_2 is the absolute value of the larger moment at the end, M_1 is the absolute value of the smaller moment at the end, for components with a single curvature, M_1/M_2 takes a positive value;

l_c——the calculated length of the component, which is approximately equal to the distance between the upper and lower support points in the corresponding principal axis direction of the eccentric compression member;

i——the cross-section radius of gyration, for the rectangular cross-section with an area of bh, $i = 0.289h$;

A——sectional area of eccentric compression members.

3. Design value of bending moment for the control section considering *P-δ* effect

Code for Design *of Concrete Structures* stipulates that except the bent structure, the eccentric compression members should take into account the axial compression force with the bending moment design values after the P-δ second-order effect. The following formulas are calculated as:

$$M = C_m \eta_{ns} M_2 \tag{5-12a}$$

$$C_m = 0.7 + 0.3 \frac{M_1}{M_2} \tag{5-12b}$$

$$\eta_{ns} = 1 + \frac{1}{1300\left(\frac{M_2}{N} + e_a\right)/h_0} \left(\frac{l_c}{h}\right)^2 \xi_c \qquad (5\text{-}12c)$$

$$\xi_c = \frac{0.5 f_c A}{N} \qquad (5\text{-}12d)$$

When $C_m \eta_{ns}$ is less than 1.0, take 1.0; for shear wall and the wall limbs of core tube, take $C_m \eta_{ns}$ equal to 1.0 for unapparent $P\text{-}\delta$ effect.

Where C_m——the section eccentricity adjustment factor for end section of the component, when less than 0.7, take 0.7;

η_{ns}——bending moment increase factor, $\eta_{ns} = 1 + \frac{\delta}{e_i}$;

e_a——additional eccentricity;

ξ_c——sectional curvature correction coefficient, when the calculated value is greater than 1.0, take 1.0;

h——calculated height; for annular cross-section, the outer diameter is taken; for circular cross-section, the diameter is taken;

h_0——the effective height for the cross-section; for the annular section, take $h_0 = r_2 + r_s$, for circular cross-section, take $h_0 = r + r_s$, where r_2 is the outer radius of the annular section, r_s is the longitudinal radius of the bar, r is the radius of the section;

A——component cross-sectional area.

5.5.2 P-δ Effect When the Directions of the Rod End Moments are Different

At this time, the components bend at double curvature, leading to a point of inflection in the member, which is common in the frame column, as shown in Figure 5-18.

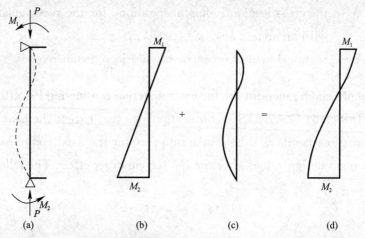

Figure 5-18 The second effect of the end moment ($P\text{-}\delta$ effect)

Although the axial force will cause additional bending moment, the bending moment at the support end is still higher than the other cross sections of the rod, that is to say, no transfer of control cross section will occur, so there is no need to consider the second-order effect in this case.

5.6 Formulae for Normal-Sectional Compressive Capacity of the Rectangular Sectional Eccentrically Compressed Members

5.6.1 Limit of Large and Small Eccentrically Compressed Failures

The basic assumptions in the calculation of the normal section in Chapter 3 are also applicable to the calculation of normal-sectional compressive bearing capacity for the eccentrically compressed member.

Similar to the bending components, the average strain distribution along the cross-section height for the normal section of eccentrically compressed member under various damage conditions can be obtained by flat section assumption and specified value of ultimate compressive strain at the edge of compression zone, as shown in Figure 5-19.

In Figure 5-19, ε_{cu} denotes the value of ultimate compressive strain of concrete at the edge of compressive zone; ε_y denotes the strain value when the tensioned longitudinal reinforcement is yielding; ε_y' denotes the strain value when the compressed longitudinal reinforcement is yielding, $\varepsilon_y' = f_y'/E_s$; x_{cb} is the height of sectional neutral axis according to strains at limit state.

As can be seen from Figure 5-19, when the height of the compression zone reaches x_{cb}, the tensioned longitudinal reinforcement reaches its yield strength. Therefore, the relative compression zone height ξ_b corresponding to the boundary damage pattern can be determined by the Eq. (3-12) in Chapter 3.

When $\xi \leqslant \xi_b$, large eccentric compression failure mode will occur; when $\xi > \xi_b$, small eccentric compression failure mode will occur.

5.6.2 Calculation of Normal-Sectional Compressive Capacity for Rectangular Sectional Eccentrically Compressed Members

1. Formula of normal-sectional compressive capacity of the rectangular section for large eccentrically compressed members

According to the simplification of bending members, the total area of compressive stress is replaced by the equivalent rectangular block, in which the stress value is taken as

$\alpha_1 f_c$, and the height of the compression zone is taken as x. Therefore, the cross-section calculation figure of large eccentrically compression failure is shown in Figure 5-20(i. e. tensile failure).

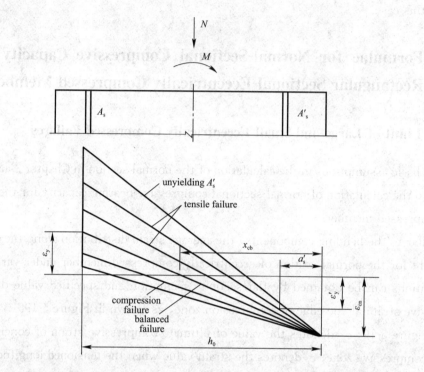

Figure 5-19 The average strain distribution along the cross section height of the eccentrically compression member in the case of various damages

(1) The equation

The following two basic equations can be obtained by the balance of force and the balance of moments caused from the tensile steel points

$$N_u = \alpha_1 f_c bx + f'_y A'_s - f_y A_s \tag{5-13}$$

$$N_u e = \alpha_1 f_c bx(h_0 - x/2) + f'_y A'_s (h_0 - a'_s) \tag{5-14}$$

$$e = e_i + h/2 - a_s \tag{5-15}$$

$$e_i = e_0 + e_a \tag{5-16}$$

$$e_0 = M/N \tag{5-17}$$

Where N_u ——design value of compressive bearing capacity;

α_1 ——coefficient, as shown in Table 3-1;

e ——the distance between axial force point to the tensioned steel bars points, as shown in Eq. (5-15);

e_i ——initial eccentricity, as calculated in Eq. (5-16);

e_0 ——the eccentricity of axial force when applied on the sectional center of gravity;

e_a —— additional eccentricity, the value of which takes the larger value of 1/30 of eccentric cross-sectional size or 20mm;

M —— design value of the bending moment in control section, the value of which is taken according to Eq. (5-12a) when considering the P-δ second-order effect;

N —— design value of axial compressive force corresponding to M;

x —— the height of the compression zone.

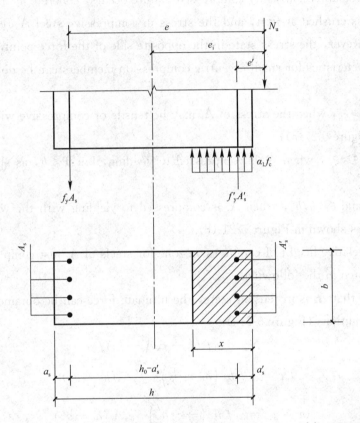

Figure 5-20　Schematic diagram of the bearing capacity of the large eccentricity compression section

(2) Applicable conditions

1) In order to ensure that the component is damaged after the tensile steel stress reaches the yield strength f_y, the height of the compression zone should meet:

$$x \leqslant x_b \tag{5-18}$$

Where　x_b —— compression zone height of concrete under boundary failure, $x_b = \xi_b h_0$, ξ_b is the same meaning with the bending member.

2) In order to ensure that stress of compressive steel can reach the yield strength f'_y when the component is damaged, the height of the compression zone should comply with

the double-bending component
$$x \geqslant 2a'_s \quad (5-19)$$

Where a'_s——distance between longitudinal compressive steel point and the edge of the compression zone.

2. Formula of normal-sectional compressive capacity of the rectangular section for small eccentrically compressive members

When the small eccentrically compressive failure occurs, concrete at the edge of compression zone is crushed at first, and the stress in compressive steel A'_s reaches the yield strength. (However, the stress state in the opposite side of the force point is unsure.)

The characteristics for small eccentric compression members can be divided into three cases:

1) $\xi_{cy} > \xi > \xi_b$, when the stress of A_s may be tensile or compressive without yielding, as shown in Figure 5-21(a);

2) $h/h_0 > \xi > \xi_{cy}$, when A_s is compressed to yielding, but $x < h$, as shown in Figure 5-21(b);

3) $\xi > \xi_{cy}$ and $\xi \geqslant h/h_0$, when A_s is compressed to yielding with the whole section in compression, as shown in Figure 5-21 (c).

ξ_{cy} is the relative height of compression zone for steels of A_s just compressed to yielding, as illustrated in the following.

Assuming that A_s is in tension state, the ultimate force can be obtained according to the balance conditions (Figure 5-21a)

$$N_u = \alpha_1 f_c bx - f'_y A'_s - \sigma_s A_s \quad (5-20)$$

$$N_u e = \alpha_1 f_c bx \left(h_0 - \frac{x}{2}\right) - f'_y A'_s (h_0 - a'_s) \quad (5-21)$$

$$\text{or } N_u e' = \alpha_1 f_c bx \left(\frac{x}{2} - a'_s\right) - \sigma_s A_s (h_0 - a'_s) \quad (5-22)$$

$$\sigma_s = \frac{\xi - \beta_1}{\xi_b - \beta_1} f_y \quad (5-23)$$

Where x——compression zone height of concrete, when $x > h$, take $x = h$;

σ_s——the stress value of the reinforcement, which can be calculated according to the flat section assumption, $-f'_y \leqslant \sigma_s \leqslant f_y$ should be met;

x_b——concrete compression zone height under the boundary failure, $x_b = \xi_b h_0$;

ξ, ξ_b——the relative compression zone height and the relative limited compression zone height, respectively;

e, e'——the distance from axial force point to the resultant point of tensile reinforcement and compressive reinforcement, respectively.

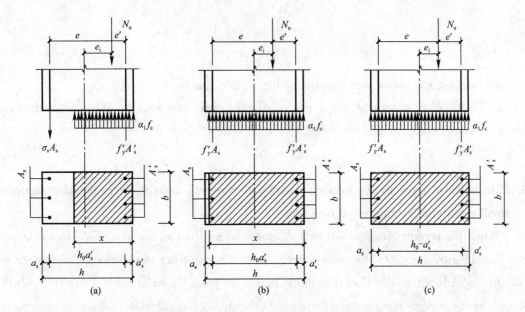

Figure 5-21 Calculated diagram of small eccentrically compressed cross-sectional bearing capacity

(a) $\xi_{cy} > \xi > \xi_b$, A_s is tensile or compressive without yielding;

(b) $h/h_0 > \xi > \xi_{cy}$, A_s is compressed to yielding, but $x < h$;

(c) $\xi > \xi_{cy}$ and $\xi \geqslant h/h_0$, A_s is compressed to yielding and all the section is compressed

$$e' = \frac{h}{2} - e_i - a'_s \tag{5-24}$$

Eq. (5-23) is explained as follows.

In the case of $x \leqslant h_0 (\xi \leqslant 1)$, the following formula can be derived by the strain diagram of Figure 5-21(a):

$$\sigma_s = \varepsilon_{cu} E_s \left(\frac{\beta_1}{\xi} - 1 \right) = \varepsilon_{cu} E_s \left(\frac{\beta_1 h_0}{x} - 1 \right) \tag{5-25}$$

Where the coefficient β_1 is the ratio between the compression zone height x of concrete and the height x_c of cross-sectional neutral axis ($x = \beta_1 \cdot x_c$). When the concrete grade $\leqslant$ C50, $\beta_1 =$ 0.8, as demonstrated in Chapter 3. However, when the stress of bar σ_s is calculated by Eq. (5-25), it is inevitable to combine the Eqs. (5-21) and (5-22) to solve the value of x, under which the cubic equation of x will appear, which is complicated in calculation. In addition, when $\xi > 1$, the deviation value of the test is large, as shown in Figure 5-22.

According to the test data in China, the measured strain of steel ε_s is almost linear with ξ, and the linear regression equation is:

$$\varepsilon_s = 0.004(0.81 - \xi) \tag{5-26a}$$

Due to the small influence on the bearing capacity of small eccentrically compressed section, considering the boundary condition that $\xi = \xi_b$, $\varepsilon_s = f_y / E_s$ and $\xi = \beta_1$, $\varepsilon_s = 0$, regression Eq. (5-26a) is simplified to the following:

$$\varepsilon_s = \frac{f_y}{E_s}\left(\frac{\beta_1 - \xi}{\beta_1 - \xi_b}\right) \tag{5-26b}$$

$$\sigma_s = \varepsilon_s E_s$$

Integrating the above two equations, Eq. (5-23) can be obtained.

In Eq. (5-23), take $\sigma_s = -f'_y$, the relative compression zone height can be obtained when A'_s is just compressed to yielding

$$\xi_{cy} = 2\beta_1 - \xi_b \tag{5-27}$$

3. Calculation of cross-section bearing capacity of reversely damage for rectangular sectional small eccentrically compressed members

When the eccentricity is small with A'_s much larger than A_s and the axial force is large, the actual centroid axis of the section deviates to A'_s, resulting in the change of the eccentric direction. It is possible that the edge concrete far from the axial force is firstly crushed, known as the reverse compression damage. At this time, the calculation diagram of sectional bearing capacity is shown in Figure 5-23.

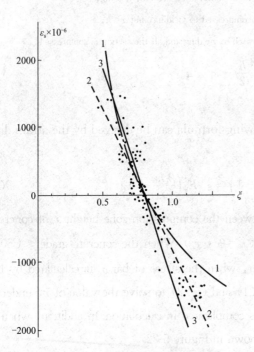

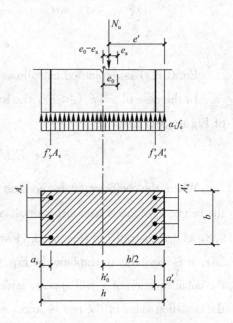

Figure 5-22 Relation curve between ε_s and ξ ($\varepsilon_{cu} = 0.0033$, $\beta_1 = 0.8$)

1—Flat section assumption of $\varepsilon_s = 0.0033\left(\frac{0.8}{\xi} - 1\right)$;

2—Regression equation of $\varepsilon_s = 0.0044(0.81 - \xi)$;

3—Simplified equation of $\varepsilon_s = \frac{f_y}{E_s}\left(\frac{0.8 - \xi}{0.8 - \xi_b}\right)$

Figure 5-23 Calculated diagram of the cross-section bearing capacity when reverse damage occurs

At this time, the additional eccentricity e_a is in reverse direction, so that e_0 decreases, namely

$$e' = \frac{h}{2} - a'_s - (e_0 - e_a) \tag{5-28}$$

Taking moment from the resultant force point of A'_s, it can be obtained:

$$A_s = \frac{N_u e' - \alpha_1 f_c bh \left(h'_0 - \frac{h}{2}\right)}{f_y (h'_0 - a_s)} \tag{5-29}$$

5.7 Calculation of Normal-Sectional Compressed Capacity for the Rectangular Sectional Eccentrically Compressed Members with Asymmetric Reinforcement

As similar with the bending capacity of the flexural members, the calculation of compressive bearing capacity of eccentrically compressed member is also divided into two types: cross-sectional design and sectional checking.

At first, it should be determined whether to consider the P-δ effect in the calculation.

5.7.1 Design of Sections with Asymmetric Reinforcement

In this calculation, the internal forces including design value N and M of component, material properties and cross-sectional size are known to obtain A_s and A'_s. The first step is to calculate the eccentricity e_i and determine the damage form of the section. For $e_i > 0.3h_0$, it belongs to the large eccentrically compressed situation; when $e_i \leqslant 0.3h_0$, it can be calculated by the case of small eccentrically compressed situation. Then the steel cross-sectional areas A_s and A'_s can be obtained by relevant formulas. The x relevant to A_s and A'_s should meet the requirements of boundary conditions. If it is not met, redesign should be conducted. In all cases, A_s and A'_s should also meet the requirement of minimum reinforcement ratio and $(A_s + A'_s)$ should not be greater than 5% of the area bh. Finally, the compressive bearing capacity perpendicular to moment action plane should be checked by the axial compressed member.

1. Design of large eccentrically compressed sections

There are two cases involved: both A_s and A'_s are unknown or A'_s is known.

(1) The cross-sectional size $b \times h$, concrete strength grade, steel types (in general, A_s and A'_s are usually taken the same kind of steel), axial force design value N and bending moment design value M and slenderness ratio l_c/h, are known: the steel cross-sectional ar-

eas A_s and A'_s are required.

Taking $N=N_u$ and $M=Ne_0$, there are three unknown variables including x, A_s and A'_s only in two equations of Eqs. (5-13) and (5-14), which is same as double-reinforced bending member. Take $x=x_b=\xi_b h_0$ in order to get the minimized total amount of reinforcement $(A_s+A'_s)$ (x_b is compression zone height) under boundary failure. Integrating $x=x_b=\xi_b h_0$ into the Eq. (5-14), the calculation equation of reinforced A'_s is

$$A'_s = \frac{Ne - \alpha_1 f_c b x_b (h_0 - 0.5 x_b)}{f'_y (h_0 - a'_s)} = \frac{Ne - \alpha_1 f_c b h_0^2 \xi_b (h_0 - 0.5\xi_b)}{f'_y (h_0 - a'_s)} \quad (5-30)$$

Take the obtained A'_s and $x=\xi_b h_0$ into Eq. (5-13)

$$A_s = \frac{\alpha_1 f_c b h_0 \xi_b - N}{f_y} + \frac{f'_y}{f_y} A'_s \quad (5-31)$$

Finally, the compressive bearing capacity perpendicular to the action plane of moment should be checked according to axial compressive component. It requires the value not less than the N value, or else redesign should be undertaken.

(2) The known variables include: b, h, N, M, f_c, f_y, f'_y, l_c/h and the number of reinforced bars A'_s and A_s is required.

Taking $N=N_u$, $M=Ne_0$, it can be seen from Eqs. (5-13) and (5-14) that there are only two unknowns x and A_s. It can directly calculate the A_s value by the joint of Eqs. (5-13) and (5-14), but it is quite complicated to solve the quadratic equation of x. Simulating from Chapter 3, taking $M_{u2}=\alpha_1 f_c b x (h_0 - x/2)$, $M_{u2}=Ne-f_y A_s(h_0-a'_s)$ is obtained by Eq. (5-14), then calculating $\alpha_s = \frac{M_{u2}}{\alpha_1 f_c b h_0^2}$ and $\xi = 1 - \sqrt{1-2\alpha_s}$ are integrated into Eq. (5-13) for A_s. It still needs to pay attention that if $x > \xi_b h_0$, recalculation should be conducted as small eccentric compression, or else the increase of the cross-sectional size, strength level of concrete, the area A'_s and other measures should be taken for large eccentric compression. If $x < 2a'_s$, according to the method of double-bouquet bending member, the moment from resultant point of compression bar A'_s is taken and the value A'_s is calculated as:

$$A_s = \frac{N\left(e_i - \frac{h}{2} + a'_s\right)}{f_y (h_0 - a')} \quad (5-32)$$

In addition, taking $A'_s=0$ as no compressed reinforcement, A_s is calculated by the Eqs. (5-13) and (5-14). The value is compared to the value calculated by Eq. (5-13), the smaller one is taken.

Finally, compressed bearing capacity perpendicular to the action plane of bending moment is checked according to axial compressive member.

It can be seen from the above that the cross-section design methods of large eccentric compression member are basically same with the double-reinforced member regardless of A'_s being unknown or known.

2. Design of small eccentrically compressed sections

In this condition, x, A_s and A_s' are unknown with only two independent balanced equations, so an extra condition should be added to solutions. It is noted that Eq. (5-23) is not a supplementary condition because of the formula $\xi = x/x_0$.

It is recommended that the cross-section design is carried out in the following two steps.

(1) Determine A_s as a supplemental condition

When $\xi_{cy} > \xi > \xi_b$, regardless of how much A_s is taken, it won't be yielded. $A_s = \rho_{min} bh = 0.002bh$ can be taken for the economy consideration, as well as taking into account the avoidance of reverse damage, so A_s can be determined by the following:

If $N \leqslant f_c bh$, $A_s = 0.002bh$

If $N > f_c bh$, $A_s = \dfrac{N_u e' - \alpha_1 f_c bh \left(h_0' - \dfrac{h}{2}\right)}{f_y (h_0' - a_s)}$; if $A_s < 0.002bh$, $A_s = 0.002bh$ is taken.

(2) Obtain the value of ξ, and then calculate A_s' with three cases of ξ

Take A_s and Eq. (5-23) into the moment balance Eq. (5-22), ξ and σ_s can be obtained. Then A_s' can be obtained according to the following three cases of small eccentric compression.

1) When $\xi_{cy} > \xi > \xi_b$, ξ and σ_s are taken into Eq. (5-20) for A_s'.

2) When $h/h_0 > \xi \geqslant \xi_{cy}$, $\sigma_s = -f_y'$ is taken, ξ is recalculated according to Eq. (5-22) by the following formula:

$$\xi = \dfrac{a_s'}{h_0} + \sqrt{\left(\dfrac{a_s'}{h_0}\right)^2 + 2\left[\dfrac{Ne'}{\alpha_1 f_c bh_0^2} - \dfrac{A_s}{bh_0} \dfrac{f_y}{\alpha_1 f_c}\left(1 - \dfrac{a_s'}{h_0}\right)\right]} \qquad (5\text{-}33)$$

And then A_s' is obtained according to Eq. (5-21).

3) When $\xi > \xi_{cy}$ and $\xi \geqslant h/h_0$, $x = h$, $\sigma_s = -f_y'$ and $\alpha_1 = 1$ are taken, A_s' can be available by the Eq. (5-21).

$$A_s' = \dfrac{Ne - f_c bh(h_0 - 0.5h)}{f_y'(h_0 - a_s')} \qquad (5\text{-}34)$$

If the value of A_s obtained above is less than $0.002bh$, it should take $A_s' = 0.002bh$.

5.7.2 Evaluation of Ultimate Capacities of Existing Sections

When reviewing bearing capacity, b, h, A_s, A_s', concrete strength grade, steel level and slenderness ratio l_c/h_0 generally are known. There are two cases: one is that the design value of axial force is known to obtain the eccentricity e_0, that is, the bending moment design value M is to be checked to withstand on the cross section; the other is that e_0 is known, the axial force design value N is to be decided. In any case, it is necessary to check the bearing capacity perpendicular to the action plane of moment.

1. Reviewing the bearing capacity of the action plane of moment

(1) Determining M when N is known

Firstly, the known reinforcement and ξ_b are taken into Eq. (5-13) to calculate compressed bearing capacity design value N_{ub} under the boundary failure. If $N \leqslant N_{ub}$, it is large eccentric compression, x is obtained according to Eq. (5-13). Then x is involved into Eq. (5-14) for e, and the design value of moment $M = Ne_0$ can be obtained. If $N > N_{ub}$, it is small eccentric compression, firstly assuming that it belongs to the first case of small eccentric compressed condition to calculate x by Eqs. (5-20) and (5-23). If $x < \xi_{cy} h_0$, the assumption is correct, taking x into Eq. (5-21) for e. e_0 can be obtained by Eq. (5-16), as well as $M = Ne_0$. If $x \geqslant \xi_{cy} h_0$, x should be obtained according to Eq. (5-33); when $x \geqslant h$, $x = h$ is taken.

Another way is to assume $\xi \leqslant \xi_b$ firstly, and x is obtained according to Eq. (5-13). If $\xi = x/h_0 \leqslant \xi_b$, the assumption is correct, and then e_0 is calculated by Eq. (5-14); if $\xi = x/h_0 > \xi_b$, the assumption is wrong that x should be based on the small eccentric compressed situation, subsequently e_0 is achieved by Eq. (5-21).

(2) Determining N when e_0 is known

x can be got by taking the moment to action point of N(Figure 5-20) because cross-section reinforcement is known.

When $x \leqslant x_b$, it is large eccentric compression, under which x and other known variables are taken into Eq. (5-13) to solve the axial force design value N. When $x > x_b$, it is small eccentric compression, under which the known variables are taken into Eqs. (5-20), (5-21) and (5-23) to solve the axial force design value N.

It can be seen from the above that when reviewing the bearing capacity of action plane of moment, as same as reviewing normal sectional bearing capacity of bending member, x is an essential factor to solve the problem.

2. Reviewing the bearing capacity perpendicular to action plane of moment

Whether it is a design or cross-section reviewing problem under large eccentric compression or small eccentric compression, besides calculations according to eccentric compression in the action plane of moment, axial compressed bearing capacity perpendicular to the action plane of moment should all be checked, and the value of φ will be considered under b as the section height.

[**Example 5-4**] Known: the design value of column's axial force is $N = 396$kN, the design values of rod end's moments are $M_1=0.92M_2$ and $M_2=220$kN·m, the cross-sectional dimension is $b=300$mm, $h=400$mm, $a_s=a'_s=40$mm; the concrete strength grade is C30 and the steel level is HRB400; $l_c/h_b=6$.

The cross-section areas of A'_s and A_s are required.

[Solution]

Because $\dfrac{M_1}{M_2}=0.92>0.9$, the P-δ effect should be considered.

$$C_m = 0.7+0.3\dfrac{M_1}{M_2}=0.976$$

$$\zeta_c = \dfrac{0.5f_cA}{N} = 0.5\times\dfrac{14.3\times 300\times 400}{396\times 10^3} = 2.17>1, \text{ take } \zeta_c = 1$$

$$e_a = 20\text{mm}$$

$$\eta_{ns} = 1+\dfrac{1}{1300\dfrac{\left(\dfrac{M_2}{N}+e_a\right)}{h_0}}\left(\dfrac{l_c}{h}\right)^2\zeta_c = 1+\dfrac{1}{1300\dfrac{\left(\dfrac{218\times 10^6}{396\times 10^3}+20\right)}{360}}6^2\times 1 = 1.017$$

$$C_m\eta_{ns} = 0.976\times 1.014 = 0.993 < 1, \text{ take } C_m\eta_{ns} = 1$$

$$M = C_m\eta_{ns}M_2 = M_2 = 220\text{kN·m}$$

So $e_i = \dfrac{M}{N}+e_a = \dfrac{220\times 10^6}{396\times 10^3}+20 = 556+20 = 576$mm

Because $e_i = 576$mm $> 0.3h_0 = 0.3\times 360 = 108$mm, so it can be calculated according to the large eccentrically compressive situation.

$$e = e_i+h/2-a_s = 576+400/2-40 = 736\text{mm}$$

From the Eq. (5-30)

$$A'_s = \dfrac{Ne-\alpha_1 f_c bh_0^2\xi_b(1-0.5\xi_b)}{f'_y(h_0-a'_s)}$$

$$= \dfrac{396\times 10^3\times 731-1.0\times 14.3\times 300\times 360^2\times 0.518\times(1-0.5\times 0.518)}{360\times(360-40)}$$

$$= 660\text{mm}^2 > \rho'_{\min}bh = 0.002\times 300\times 400 = 240\text{mm}^2$$

From the Eq. (5-31)

$$A_s = \dfrac{\alpha_1 f_c bh_0\xi_b-N}{f_y}+\dfrac{f'_y}{f_y}A'_s = \dfrac{1.0\times 14.3\times 300\times 360\times 0.518-396\times 10^3}{360}+660$$

$$= 1782\text{mm}^2$$

Tensioned steels A_s are adopted as 3⌀22+2⌀20 ($A_s=1768$mm^2), compressed steels A'_s are adopted as 2⌀18+1⌀14 ($A'_s=662.9$mm^2).

According to Eq. (5-13), x is figured out:

$$x = \dfrac{N-f'_yA'_s+f_yA_s}{\alpha_1 f_c b} = \dfrac{396\times 10^3-360\times 662.9+360\times 1768}{1.0\times 14.3\times 300} = 185\text{mm}^2$$

$\xi = \dfrac{x}{h_0} = \dfrac{185}{360} = 0.514 < \xi_0 = 0.518$, the assumption of large eccentric compression is correct.

The bearing capacity perpendicular to the action plane of moment satisfies the requirements.

[**Example 5-5**] The conditions are the same with Example 5-4, and $A'_s = 882\text{mm}^2$ (1 ⌽ 18+2 ⌽ 20) is known.

Tensioned steels' cross-section area A_s is required.

[Solution]

$N = N_u$ is taken.

From Eq. (5-14)

$M_{u2} = Ne - f'_y A'_s (h_0 - a'_s) = 396 \times 10^3 \times 736 - 360 \times 882 \times (360 - 40) = 190\text{kN} \cdot \text{m}$

$$\alpha_s = \dfrac{M_{u2}}{\alpha_1 f_c b h_0^2} = \dfrac{190 \times 10^6}{1 \times 14.3 \times 300 \times 360^2} = 0.342$$

$\zeta = 1 - \sqrt{1 - 2\alpha_s} = 1 - \sqrt{1 - 2 \times 0.342} = 0.438 < \xi_b = 0.518$, it belongs to large eccentric compression member.

$$x = \xi h_0 = 0.438 \times 360 = 158\text{mm} > 2a'_s = 2 \times 40 = 80\text{mm}$$

From the Eq. (5-28)

$A_s = \dfrac{\alpha_1 f_c b x + f'_y A'_s - N}{f_y} = \dfrac{1 \times 14.3 \times 300 \times 158 + 360 \times 882 - 396 \times 10^3}{360} = 1665\text{mm}^2$

2 ⌽ 20+2 ⌽ 25 is used ($A_s = 1610\text{mm}^2$).

It can be seen that comparing Example 5-4 with Example 5-5, when $\xi = \xi_b$ is adopted, the calculation value of total amount of steel is $660 + 1782 = 2442\text{mm}^2$, which is less than the total amount of steel obtained from Example 5-5 as $882 + 1665 = 2547\text{mm}^2$.

[**Example 5-6**] $N = 160\text{kN}$, rod end's design value is $M_1 = M_2 = 250.9\text{kN} \cdot \text{m}$, $b = 300\text{mm}$, $h = 500\text{mm}$, $a_s = a'_s = 40\text{mm}$. Compressed steels are used as 4 ⌽ 22 of $A'_s = 1520\text{mm}^2$ (HRB400), concrete strength grade is C30 and the calculated length is $l_c = 6\text{m}$.

The sectional area A_s is required.

[Solution]

From Eq. (5-11c)

$$\dfrac{l_c}{i} = \dfrac{l_c}{\sqrt{\dfrac{1}{12}}h} = \dfrac{6000}{0.289 \times 500} = 41.5 > 34 - 12\left(\dfrac{M_1}{M_2}\right) = 22$$

Thus the P-δ effect should be considered:

$$\dfrac{M_2}{N} = \dfrac{250.9 \times 10^3}{160 \times 10^3} = 1568\text{mm}$$

$$e_0 = 20\text{mm}$$

$$\xi_c = \dfrac{0.5 f_c A}{N} = \dfrac{0.5 \times 14.3 \times 300 \times 500}{160 \times 10^3} = 6.70 > 1,\ \zeta_c = 1$$

$$\frac{l_c}{h} = \frac{6000}{500} = 12$$

$$\eta_{ns} = 1 + \frac{1}{\frac{1300(M_2/N+e_a)}{h_0}} \left(\frac{l_c}{h}\right)^2 \zeta_c$$

$$= 1 + \frac{1}{1300 \times 3.452} \times 12^2 \times 1 = 1.032$$

$$C_m = 0.7 + 0.3 \frac{M_1}{M_2} = 1$$

$$M = C_m \eta_{ns} M_2 = 1 \times 1.032 \times 250.9 = 259 \text{kN} \cdot \text{m}$$

$e_i = \frac{M}{N} + e_0 = 1619 + 20 = 1639\text{mm} > 0.3h_0 = 0.3 \times 460 = 138\text{mm}$, it can be calculated by large eccentrically compressive situation.

$$e = e_i + \frac{h}{2} - a_s = 1639 + 500/2 - 40 = 1849\text{mm}$$

$$M_{u2} = Ne - f'_y A'_s (h_0 - a'_s)$$
$$= 160 \times 10^3 \times 1849 - 360 \times 1520 \times (460 - 40)$$
$$= 66.02 \text{kN} \cdot \text{m}$$

$$\alpha_s = \frac{M_{u2}}{\alpha_1 f_c b h_0^2} = \frac{66.02 \times 10^6}{1 \times 14.3 \times 300 \times 460^2} = 0.073$$

$\xi = 1 - \sqrt{1-2\alpha_s} = 1 - \sqrt{1-2 \times 0.073} = 0.073 < \xi_b = 0.518$, which proves that assumption of large eccentric compression is correct.

$$x = \xi h_0 = 0.076 \times 460 = 35\text{mm} < 2a'_s = 80\text{mm}$$

A_s is obtained by Eq. (5-32)

$$A_s = \frac{N \cdot (e_i - h/2 + a'_s)}{f_y(h_0 - a'_s)} = \frac{160 \times 10^3 \times (1639 - 500/2 + 40)}{360 \times (460 - 40)} = 1512\text{mm}^2$$

For situation without considering the performance of A'_s

$$M_{u2} = Ne = 160 \times 10^3 \times 1849 = 295.84 \text{kN} \cdot \text{m}$$

$$\alpha_s = 0.326, \ \xi = 0.410, \ x = 189\text{mm}, \ A_s = 3327\text{mm}^2$$

All above proves that tensioned steel will get a higher value without compressive steels. The answer to this exercise is $A_s = 1512\text{mm}^2$ of selecting 4 ⏀ 22 ($A_s = 1512\text{mm}^2$).

[**Example 5-7**] Known: the design value of column's axial force under load is $N = 3100\text{kN}$, the design values of rod end's moment are $M_1 = 0.95M_2$, $M_2 = 85\text{kN} \cdot \text{m}$; the cross-sectional dimension is $b = 400\text{mm}$, $h = 600\text{mm}$, $a_s = a'_s = 40\text{mm}$; strength grade of concrete is C40. Stirrups are HPB400. $A'_s = 1964\text{mm}^2$ (4 ⏀ 25), $A_s = 603\text{mm}^2$ (3 ⏀ 16). The calculation length of column l_0 is 6m. M_u is to be determined.

[**Solution**]

$$f_c = 19.1\text{N/mm}^2, \ f_y = f'_y = 360\text{N/mm}^2$$

$$\xi_b = 0.518, \alpha_1 = 1.0$$

$$x = \frac{N - f'_y A'_s + f_y A_s}{\alpha_1 f_c b}$$

$$= \frac{3100 \times 10^3 - 1964 \times 360 + 603 \times 360}{1.0 \times 19.1 \times 400} = 341 \text{mm} > \xi_b h_0 = 290 \text{mm}$$

So, it belongs to the condition of small eccentric compression.

Then, look up in Tab. 5-1, $\frac{l_0}{b} = 15$, $\varphi = 0.895$

$$N_u = 0.9\varphi[f_c A + f'_y(A'_s + A_s)]$$
$$= 0.9 \times 0.895 \times [19.1 \times 400 \times 600 + 360 \times (1964 + 603)] = 4437 \text{kN} > N$$

Recalculate x:

Due to
$$N_u = \alpha_1 f_c b x + f'_y A'_s - \sigma_s A_s, \quad \sigma_s = \frac{\xi - \beta_1}{\xi_b - \beta_1} f_y$$

for
$$N_u = \alpha_1 f_c b x + f'_y A'_s - \frac{\xi - \beta_1}{\xi_b - \beta_1} f_y A_s,$$

$$3100 \times 10^3 = 1.0 \times 19.1 \times 400 \times 560\xi + 1964 \times 360 - \frac{\xi - 0.8}{0.518 - 0.8} \times 360 \times 603$$

$$\xi = 0.596$$

$$x = \xi h_0 = 334 \text{mm}$$

Due to
$$Ne = \alpha_1 f_c b x \left(h_0 - \frac{x}{2} \right) + f'_y A'_s (h_0 - a'_s)$$

So
$$e = \frac{\alpha_1 f_c b x \left(h_0 - \frac{x}{2} \right) + f'_y A'_s (h_0 - a'_s)}{N}$$

$$= \frac{1.0 \times 19.1 \times 400 \times 334 \times \left(560 - \frac{334}{2} \right) + 1964 \times 360 \times (560 - 40)}{3100 \times 10^3} = 442 \text{mm}$$

$$e' = e - \frac{h}{2} + a_s = 442 - \frac{600}{2} + 40 = 182 \text{mm}$$

$$e_a = 20 \text{mm}$$

$$e_0 = 162 \text{mm}$$

$$M_u = Ne_0 = 3100 \times 10^3 \times 162 = 502.2 \text{kN} \cdot \text{m}$$

Due to $\frac{M_1}{M_2} = 0.95 > 0.9$, the P-δ effect should be considered.

$$C_m = 0.7 + 0.3 \frac{M_1}{M_2} = 0.7 + 0.3 \times 0.95 = 0.985$$

$$\zeta_c = 0.5 \frac{f_c A}{N} = 0.5 \times \frac{19.1 \times 400 \times 600}{3100 \times 10^3} = 0.739$$

$$\eta_{ns} = 1 + \frac{1}{1300 \left(\frac{M_2}{N} + e_a \right)/h_0} \left(\frac{l_c}{h} \right)^2 \zeta_c$$

$$= 1 + \frac{1}{1300\left(\frac{85 \times 10^6}{3100 \times 10^3} + 20\right)/560} \left(\frac{6 \times 10^3}{600}\right)^2 \times 0.739 = 1.671$$

$$C_m \eta_{ns} = 0.985 \times 1.671 = 1.646 > 1.0$$

So, $M = C_m \eta_{ns} M_2 = 1.646 \times 85 = 140 \text{kN} \cdot \text{m}$

$$M_u > M$$

Hence the column is safe.

[Example 5-8] $N=1200$kN, $b=400$mm, $h=600$mm, $a_s=a'_s=40$mm, the strength grade of concrete is C40 and HRB400 is used. A_s is taken as 4⏀20 and A'_s is adopted as 4⏀20 ($A_s=1256$mm², $A'_s=1520$mm²). The calculated length of member is 4m. The ratio between design values of the rod ends' moments is $M_1 = 0.85 M_2$.

The design value of bending moment for the section in the direction of h is required.

[Solution]

Because $\frac{M_1}{M_2} = 0.85 < 0.9$

$$\frac{N}{f_c A} = 0.26 < 0.9$$

$$\frac{l_c}{i} = 23.1 < 34 - 12\left(\frac{M_1}{M_2}\right) = 23.8$$

Thus, the effect of P-δ is not taken into account.

$N = N_u$, according to Eq. (5-13)

$$x = \frac{N - f'_y A'_s + f_y A_s}{\alpha_1 f_c b}$$

$$= \frac{1200 \times 10^3 - 360 \times 1520 + 360 \times 1256}{1.0 \times 19.1 \times 400}$$

$$= 145 \text{mm} < \xi_b h_0 (= 0.518 \times 560 = 290 \text{mm})$$

It belongs to the condition of large eccentric compression. $x = 145$mm $> 2a'_s (= 2 \times 40 = 80$mm), indicating that the compressive reinforcement can reach the yielding strength. According to Eq. (5-14).

$$e = \frac{\alpha_1 f_c bx \left(h_0 - \frac{x}{2}\right) + f'_y A'_s (h_0 - a'_s)}{N}$$

$$= \frac{1.0 \times 19.1 \times 400 \times 145 \times \left(560 - \frac{145}{2}\right) + 360 \times 1520 \times (560 - 40)}{1200 \times 10^3} = 687 \text{mm}$$

$$e_i = e - \frac{h}{2} + a_s = 687 - \frac{600}{2} + 40 = 427 \text{mm}$$

Due to $e_i = e_0 + e_a$, $e_a = 20$mm

$$e_0 = e_i - e_a = 427 - 20 = 407 \text{mm}$$

$$M = Ne_0 = 1200 \times 0.407 = 488.4 \text{kN} \cdot \text{m}$$

The design value of bending moment for the section can be borne in h direction.

$$M = 488.4 \text{kN} \cdot \text{m}$$

[**Example 5-9**] The sectional size of frame column is $b=500$mm, $h=700$mm and $a_s = a'_s = 45$mm. The strength level of concrete is C35 and the steel bars are HPB400. $A_s = 2281$mm^2, $A'_s = 1520$mm^2, $l_c = 12.25$m and $e_0 = 600$mm.

N_u is to be determined.

[**Solution**]

The point of inflection for the frame column is within the column so that P-δ effect is not to be considered.

$$e_0 = 600\text{mm}, \quad e_a = 700/30 = 23\text{mm}(>20\text{mm})$$

Thus, $e_i = e_0 + e_a = 600 + 23 = 623$mm

$$\alpha_1 f_c bx \left(e_i - \frac{h}{2} + \frac{x}{2}\right) = f_y A_s \left(e_i + \frac{h}{2} - a_s\right) - f'_y A'_s \left(e_i - \frac{h}{2} + a'_s\right)$$

Taking the data into the above equation

$$1.0 \times 16.7 \times 500 \times x \times \left(623 - 350 + \frac{x}{2}\right)$$

$$= 360 \times 2281 \times (623 + 350 - 45) - 360 \times 1520 \times (623 - 350 + 45)$$

Solving the above equation, $x^2 = 564x - 140844 = 0$

$$x = \frac{1}{2} \times (-546 + \sqrt{546^2 + 4 \times 140844}) = 191\text{mm}$$

$2a'_s(= 2 \times 45 = 90\text{mm}) < x < x_b(= 0.518 \times 655 = 339\text{mm})$

$N_u = \alpha_1 f_c bx + f'_y A'_s - f_y A_s$

$= 1.0 \times 16.7 \times 500 \times 191 + 360 \times 1520 - 360 \times 2281 = 1320.9$kN $\cdot$ m

The design value of the axial force for the section is

$$N_u = 1320.9\text{kN}$$

5.8 Calculation of Normal-Sectional Compressive Capacity for the Rectangular Sectional Eccentrically Compressed Members with Symmetric Reinforcement

In practical engineering, the eccentrically compressed member may suffer bending moments from the opposite directions under different loading conditions.

Symmetric reinforcement ought to be used when bending moment from opposite direction has great difference, but the total amount of longitudinal reinforcement in symmetric reinforcement design increases little comparing with the total amount of longitudinal reinforcement in unsymmetric reinforcement design. Fabricated columns generally use symmetrical reinforcement in order to prevent lifting errors.

5.8.1 Design of Sections with Symmetric Reinforcement

When it is the symmetrical reinforcement, the sectional reinforcement on both sides are the same, namely $A'_s = A_s$, $f_y = f'_y$.

1. Calculation of large eccentrically compressed members

$N = N_u$, according to Eq. (5-13)

$$x = \frac{N}{\alpha_1 f_c b} \tag{5-35}$$

Integrating in Eq. (5-14), the solution is:

$$A_s = A'_s = \frac{Ne - \alpha_1 f_c bx\left(h_0 - \frac{x}{2}\right)}{f'_y(h_0 - a'_s)} \tag{5-36}$$

When $x < 2a'_s$, asymmetric reinforcement should be conducted. If $x > x_b$, the tensioned bars A_s cannot reach the yielding strength, leading to the situation of compressed damage. Under that situation, the small eccentrically compressed formula should be undertaken.

2. Calculation of the small eccentrically flexural members

Because of the symmetric reinforcement, $A_s = A'_s$. Thus x and $A_s = A'_s$ can be directly calculated by Eqs. (5-20), (5-21) and (5-23). Integrating $f_y = f'_y$, $x = \xi/h_0$, $N = N_0$ and Eq. (5-23) into Eq. (5-20), the following can be obtained:

$$N = \alpha_1 f_c b h_0 \xi + (f'_y - \sigma_s) A'_s$$

Also

$$f'_y A'_s = \frac{N - \alpha_1 f_c b h_0 \xi}{\frac{\xi_b - \xi}{\xi_b - \beta_1}}$$

Taken into Eq. (5-21)

$$Ne = \alpha_1 f_c b h_0^2 \xi\left(1 - \frac{\xi}{2}\right) + \frac{N - \alpha_1 f_c b h_0 \xi}{\frac{\xi_b - \xi}{\xi_b - \beta_1}}(h_0 - a'_s)$$

Also

$$Ne\frac{\xi_b - \xi}{\xi_b - \beta_1} = \alpha_1 f_c b h_0 \xi(1 - 0.5\xi)\left(\frac{\xi_b - \xi}{\xi_b - \beta_1}\right) + (N - \alpha_1 f_c b h_0 \xi) \times (h_0 - a'_s) \tag{5-37}$$

By Eq. (5-37), it's hard to calculate the cubic equation of $x(x = \xi h_0)$ manually. Simplification method should be conducted as follows.

Taking

$$\overline{y} = \xi(1 - 0.5\xi)\left(\frac{\xi_b - \xi}{\xi_b - \beta_1}\right) \tag{5-38}$$

Plug in Eq. (5-37):

$$\frac{Ne}{\alpha_1 f_c b h_0^2}\left(\frac{\xi_b - \xi}{\xi_b - \beta_1}\right) - \left(\frac{Ne}{\alpha_1 f_c b h_0^2} - \xi/h_0\right)(h_0 - a_s') = \overline{y} \quad (5\text{-}39)$$

ξ_b and β_1 are known for given steel grade and concrete strength grade. We can draw relation curve between $\overline{y}$ and ξ according to Eq. (5-39) (Figure 5-24).

According to Figure 5-24, the relation between $\overline{y}$ and ξ is similar to a straight line during the section of small eccentric compression. For the steel of HPB300, HRB335, HRB400 (or RRB400), the equation of $\overline{y}$ and ξ is similar to

$$\overline{y} = 0.43 \frac{\xi - \xi_b}{\beta_1 - \xi_b} h \quad (5\text{-}40)$$

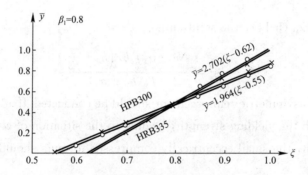

Figure 5-24　Relation curve between $\overline{y}$ and ξ

Taking Eq. (5-40) into (5-39), the approximate formula for ξ given by *Code for Design of Concrete Structures* is obtained after integrating process.

$$\xi = \frac{N - \xi_b \alpha_1 f_c b h_0}{\dfrac{N - 0.43\alpha_1 f_c b h_0^2}{(\beta_1 - \xi_b)(h_0 - a_s')} + \alpha_1 f_c b h_0} + \xi_b \quad (5\text{-}41)$$

Plug in Eq. (5-36), the reinforcement area is:

$$A_s' = A_s' = \frac{Ne - \alpha_1 f_c b h_0^2 \xi(1 - 0.5\xi)}{f_y'(h_0 - a_s')} \quad (5\text{-}42)$$

5.8.2　Evaluation of Ultimate Capacities of Existing Sections

Similar procedures can be undertaken according to the section reviewing method of asymmetric reinforcement, but $A_s = A_s'$ and $f_y = f_y'$ should be taken.

[**Example 5-10**]　The conditions are the same with Example 5-4 with symmetrical reinforcement.

The $A_s = A_s'$ is required.

[**Solution**]

According to the results in Example 5-4, $e_i = 571$mm $> 0.3h_0$, belonging to large eccentrically compressed situation.

According to Eqs. (5-35) and (5-36)

$$x = \frac{N}{\alpha_1 f_c b} = \frac{396 \times 10^3}{1.0 \times 14.3 \times 300} = 92.3 \text{mm}$$

$$2a'_s < x < 0.518 h_0$$

$$A_s = A'_s = \frac{Ne - \alpha_1 f_c bx(h_0 - 0.5x)}{f'_y(h_0 - a'_s)}$$

$$= \frac{396 \times 10^3 \times 736 - 1.0 \times 14.3 \times 300 \times 92.3 \times (360 - 92.3/2)}{360 \times (360 - 40)}$$

$$= 1451 \text{mm}^2$$

The steels are configured on each side with $3 \oplus 20 + 2 \oplus 18 (A'_s = A_s = 1451 \text{mm}^2)$.

The results in Example 5-4: $A_s + A'_s = 1782 + 662.9 = 2444.9 \text{mm}^2$. In this example, $A'_s + A_s = 2 \times 1434 = 2868 \text{mm}^2$. Compared with Example 5-4, the amount of steel increases when adopting symmetrical reinforcement.

[**Example 5-11**] The design value of axial force $N = 3800 \text{kN}$, bending moment $M_1 = 0.88 M_2$, $M_2 = 340 \text{kN} \cdot \text{m}$. Section size is $b = 400 \text{mm}$, $h = 600 \text{mm}$ and $a_s = a'_s = 45 \text{mm}$; The strength grade of concrete is C40. Steel bars are HRB400. The calculated length of component is $i_c = i_0 = 3.3 \text{m}$.

$A'_s = A_s$ is required.

[**Solution**]

$$M_1/M_2 = 0.88 < 0.9$$

$$\frac{N}{f_c A} = \frac{3500 \times 10^3}{19.1 \times 400 \times 700} = 0.65 < 0.9$$

$$\frac{l_c}{i} = \frac{3300}{0.289 \times 700} = 16.3 < 34 - 12 \frac{M_1}{M_2} = 23.4$$

Therefore, the P-δ effect should be considered.

$M = M_2 = 340 \text{kN} \cdot \text{m}$

$e_a = 700/30 = 23 \text{mm} > 20 \text{mm}$

$e_0 = M/N = (340 \times 10^6) \div (3800 \times 10^3) = 89 \text{mm}$

$e_i = e_0 + e_a = 89 + 23 = 112 \text{mm}$

$e_i = 112 \text{mm} < 0.3 h_0 = 0.3 \times 655 = 196.5 \text{mm}$

$e = e_i + h/2 - a_s = 112 + 700/2 - 45 = 417 \text{mm}$

$$x = \frac{N}{\alpha_1 f_c b} = \frac{380 \times 10^4}{1.0 \times 19.1 \times 400} = 497 \text{mm} > x_b = 0.518 \times 655 = 339 \text{mm}$$

It should be calculated according to small eccentric compression.

Due to $\beta_1 = 0.8$ and Eq. (5-41)

$$\xi = \frac{N - \xi_b \alpha_1 f_c b h_0}{\dfrac{Ne - 0.43\alpha_1 f_c b h_0^2}{(\beta_1 - \xi_b)(h_0 - a_s')} + \alpha_1 f_c b h_0} + \xi_b$$

$$= \frac{3800 \times 10^3 - 0.518 \times 1.0 \times 19.1 \times 400 \times 655}{\dfrac{3800 \times 10^3 \times 417 - 0.43 \times 1.0 \times 19.1 \times 400 \times 655^2}{(0.8 - 0.518) \times (655 - 45)} + 1.0 \times 19.1 \times 400 \times 655} + 0.518$$

$$= 0.519$$

$$x = \xi h_0 = 0.519 \times 655 = 340 \text{mm}$$

$$A_s = A_s' = \frac{Ne - \alpha_1 f_c b x \left(h_0 - \dfrac{x}{2}\right)}{f_y'(h_0 - a_s')}$$

$$= \frac{3800 \times 10^3 \times 417 - 1.0 \times 19.1 \times 400 \times 340 \times \left(655 - \dfrac{340}{2}\right)}{360 \times (655 - 45)}$$

$$= 1496 \text{mm}^2 > \rho_{\min}' bh = 0.2\% \times 400 \times 700 = 560 \text{mm}^2$$

The configuration on each side is 3 $\oplus$ 25 with $A_s' = A_s = 1473 \text{mm}^2$.

Besides, bearing capacity perpendicular to the action direction of bending moment according to axial compression should be checked.

Due to $\dfrac{l_0}{b} = \dfrac{3300}{400} = 8.25$, looking up Table 5-1, $\varphi = 0.998$.

According to Eq. (5-4)

$$N = 0.9\varphi[f_c bh + f_y'(A_s' + A_s)]$$

$$= 0.9 \times 0.998 \times [19.1 \times 400 \times 700 + 360 \times (1473 + 1473)]$$

$$= 5756 \text{kN} > 3500 \text{kN}$$

The result is under security.

In summary, when calculating cross-sectional compressed capacity of rectangular sectional eccentrically compressed members, only two equilibrium equations of force and moment are involved. Therefore, when the number of unknowns are more than two, the supplementary condition should be used (approximate calculated equation of σ_s in Eq. (5-23) for small eccentric compression also contains the unknown x, so it is not a supplemental condition). When unknowns are not more than two, the calculation should also adopt the appropriate method to solve successfully.

5.9 Calculation of Normal-Sectional Compressed Capacity of the I-shaped Sectional Eccentrically Compressed Members with Symmetric Reinforcement

In order to save concrete and reduce the weight of the member, I-shaped column is usually used for prefabricated column of larger size. The form of damage for the I-shaped sectional column is the same as rectangular section.

5.9.1 Calculation of Large Eccentric Compression Sections

1. Formulae

(1) when $x > h'_f$, the compression zone is T-section, as shown in Figure 5-25(a), the calculations are conducted as follows.

$$N_u = \alpha_1 f_c [bx + (b'_f - b)h'_f] \tag{5-43}$$

$$N_u e = \alpha_1 f_c \left[bx \left(h_0 - \frac{x}{2}\right) + (b'_f - b)h'_f \left(h_0 - \frac{h'_f}{2}\right) \right] + f'_y A'_s (h_0 - a'_s) \tag{5-44}$$

(2) When $x \leqslant h'_f$, according to rectangular section of width b'_f in Figure 5-25(b).

$$N_u = \alpha_1 f_c b'_f x \tag{5-45}$$

$$N_u e = \alpha_1 f_c b'_f x \left(h_0 - \frac{x}{2}\right) + f'_y A'_s (h_0 - a'_s) \tag{5-46}$$

Where b'_f——compressed flange width of I-shaped section;

        h'_f——compressed flange height of I-shaped section.

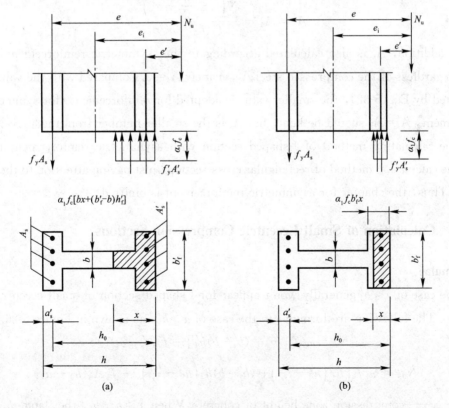

Figure 5-25 The calculated diagram of I-shaped sectional large eccentric compression
(a)$x > h'_f$; (b)$x \leqslant h'_f$;

2. Applicable conditions

In order to ensure that the tensile steel bar A_s and the compressed bar A_s' of above calculation formulas can reach the yielding strength, the following conditions should be met.

$$x \leqslant x_b \text{ and } x \geqslant 2a_s'$$

Where x_b is the calculation height of compression zone under boundary failure.

3. Calculation method

I-shaped section is assumed to a rectangular section with a width b_f', from Eq. (5-45):

$$x = \frac{N_u}{\alpha_1 f_c b_f'} \qquad (5\text{-}47)$$

There are three cases according to the different values of x:

1) When $x > h_f'$, the sectional area of steel bar can be obtained by using Eqs. (5-43) and (5-44). At the same time, $x \leqslant x_b$ should be met;

2) When $2a_s' \leqslant x \leqslant h_f'$, the sectional area of steel bar can be obtained by using Eq. (5-46).

3) When $x < 2a_s'$, as the same with double-bar bending component, $x = 2a_s'$ is taken, and reinforcement can be obtained according to Eq. (5-32)

$$A_s' = A_s = \frac{N\left(e_i - \frac{h}{2} + a_s'\right)}{f_y(h_0 - a_s')}$$

In addition, A_s is also calculated according to the asymmetric reinforcement components regardless of the compressed steel A_s', namely $A_s' = 0$. Compared with the value of A_s calculated by Eq. (5-32), the smaller value is adopted for reinforcement (for symmetric reinforcement, $A_s' = A_s$ should be kept, but A_s is the smaller number from the above).

The calculating method of I-shaped section with asymmetric reinforcement is same with the calculation method of rectangular cross-section, just paying attention to the role of flange. Thus, the chapter for asymmetric reinforcement is omitted.

5.9.2 Calculation of Small Eccentric Compression Sections

1. Formulae

The case of $x < h_f'$ generally won't appear for I-shaped section of small eccentric compression. The following are formulas in the case of $x > h_f'$, as shown in Figure 5-26.

$$N_u = \alpha_1 f_c [bx + (b_f' - b)h_f'] + f_y' A_s' - \sigma_s A_s \qquad (5\text{-}48)$$

$$N_u e = \alpha_1 f_c \left[bx \left(h_0 - \frac{x}{2}\right) + (b_f' - b)h_f' \left(h_0 - \frac{h_f'}{2}\right) \right] + f_y' A_s' (h_0 - a_s') \qquad (5\text{-}49)$$

Where x——compression zone height of concrete. When $x > h - h_f$, the flange h_f should be considered in the calculation, which is calculated according to Eqs. (5-50) and (5-51).

$$N_u = \alpha_1 f_c [bx + (b'_f - b)h'_f + (b_f - b)(h_f + x - h)] + f'_y A'_s - \sigma_s A_s \quad (5\text{-}50)$$

$$N_u e = \alpha_1 f_c \left[bx \left(h_0 - \frac{x}{2} \right) + (b'_f - b)h'_f \left(h_0 - \frac{h'_f}{2} \right) + (b_f - b)(h_f + x - h) \left(h_f - \frac{h_f + x - h}{2} - a_s \right) \right]$$
$$+ f'_y A'_s (h_0 - a'_s) \quad (5\text{-}51)$$

When x is greater than h in the formula, take $x = h$. σ_s can still be calculated by Eq. (5-23).

For the small eccentrically compressed members, the following conditions should be met.

$$N_u \left[\frac{h}{2} - a'_s - (e_0 - e_a) \right]$$
$$\leqslant \alpha_1 f_c \left[bh \left(h'_0 - \frac{h}{2} \right) + (b_f - b)h_f \left(h'_0 - \frac{h_f}{2} \right) + (b'_f - b)h'_f \left(\frac{h'_f}{2} - a'_s \right) \right] \quad (5\text{-}52)$$
$$+ f'_y A_s (h'_0 - a'_s)$$

Where h'_0——the distance between the resultant point of A'_s and the edge far from the longitudinal force N, $h'_0 = h - a'_s$.

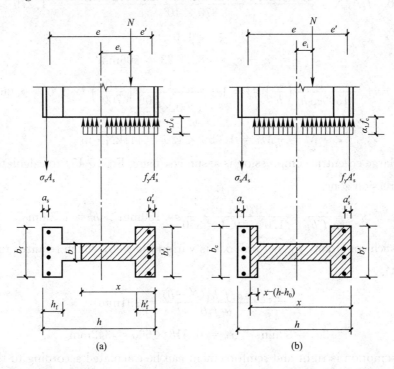

Figure 5-26 Calculation of small eccentric compression with I-shaped section
(a) $\xi_b h_0 < x < h - h_f$; (b) $h - h_f < x < h$

2. Applicable Conditions

The applicable condition for small eccentric compression with I-shaped section is $x > x_b$.

3. Calculation Method

The calculation method of I-shaped section with symmetrical reinforcement is basically the same with rectangular section of symmetrical reinforcement. Generally, iterative method and approximate formula method can be used. When using iterative method, σ_s is still calculated by Eq. (5-23). Eqs. (5-20) and (5-21) are replaced respectively by Eq. (5-48) and Eq. (5-49) or Eq. (5-50) and Eq. (5-51), as shown in the examples below.

[**Example 5-12**] The side pillar of I-shaped section with $l_0 = l_c = 5.7$m, the controlling internal force of pillar's section are $N = 870$kN and $M_1 = 0.95 M_2$ with $M_2 = 420$kN·m. Section size is $b = 80$mm, $h = 700$mm, $b_f = b'_f = 350$mm, $h_f = h'_f = 112$mm and $a_s = a'_s = 45$mm; The strength grade of concrete is C40. Steel bars are HRB400. Symmetrical reinforcement $A'_s = A_s$ is required.

[**Solution**]

$$f_c = 19.1 \text{N/mm}^2, \ f_y = f'_y = 360 \text{N/mm}^2$$

$$\xi_b = 0.518, \ \alpha_1 = 1.0$$

$$\xi_c = 0.5 f_c A/N = \frac{0.5 \times 19.1 \times (80 \times 700 + 2 \times 112 \times 270)}{870 \times 10^3} = 1.279 > 1.0$$

$$\xi_c = 1.0$$

$$e_i = e_0 + e_a = 483 + 23 = 506 \text{mm}$$

$$\eta_s = 1 + \frac{1}{1300 e_i/h_0} \left(\frac{l_0}{h}\right)^2 \xi_c = 1 + \frac{1}{1300 \times \frac{506}{660}} \left(\frac{5700}{700}\right)^2 \times 1.0 = 1.067$$

$$M = \eta_s M_0 = 1.058 \times 420 = 441 \text{kN} \cdot \text{m}$$

Firstly large eccentric compression is assumed, using Eq. (5-45) to calculate the height of the compression zone.

$$x = \frac{N}{\alpha_1 f_c b'_f} = \frac{870 \times 10^3}{1.0 \times 19.1 \times 350} = 130 \text{mm} > h'_f = 112 \text{mm}$$

Under such condition, the netural axis is within the web, which requires to recalculate by Eq. (5-43).

$$x = \frac{N - \alpha_1 f_c h'_f (b'_f - b)}{\alpha_1 f_c b} = 191 \text{mm}$$

$$2 a'_s = 80 \text{mm} < x_b = 0.518 \times 660 = 342 \text{mm}$$

The assumption is right and reinforcement can be calculated according to the formulas of large eccentric compression.

$$e_i = e_0 + e_a = \frac{441 \times 10^6}{870 \times 10^3} + 23 = 507 \text{mm}$$

$$e = e_i + \frac{h}{2} - a_s = 533 + 350 - 40 = 817 \text{mm}$$

$$A'_s = \frac{Ne - \alpha_1 f_c \left[bx \left(h_0 - \frac{x}{2} \right) + (b'_f - b) h'_f \left(h_0 - \frac{h'_f}{2} \right) \right]}{f'_y (h_0 - a'_s)}$$

$$= \frac{870 \times 10^3 \times 817 - 1.0 \times 19.1 \left[80 \times 191 \times \left(660 - \frac{191}{2}\right) + (350 - 80) \times 112 \times \left(660 - \frac{112}{2}\right)\right]}{360 \times (660 - 40)}$$

$= 985 \text{mm}^2 > \rho'_{\min} A = 0.002 \times (700 \times 80 + 2 \times 270 \times 112) = 233 \text{mm}^2$

Each side is chosen as 2 Φ 25 with $A'_s = A_s = 982 \text{mm}^2$.

5.10　N_u-M_u Interaction Diagram and Its Application

The experiments illustrate that under small eccentric compression, the flexural bearing capacity of normal section decreases with the increasing of axial force while under large eccentric compression, the trend is on the contrary. Under boundary damage, the flexural bearing capacity of normal section reaches the maximum.

Figure 5-27 is a test curve between M_u and N_u under different eccentricities according to a group of eccentric compression specimens conducted by Southwest Jiaotong University. The curve reflects the above rules.

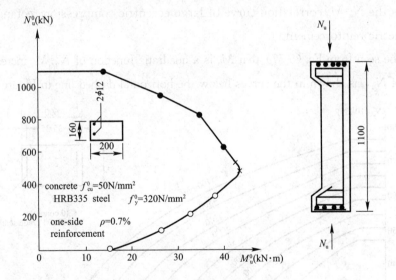

Figure 5-27　Test related curve of N_u-M_u

It illustrates that for the eccentric compression member with a given section size, the reinforcement and material strength can reach the limited bearing capacity in a series of different combinations between M_u and N_u. In other words, for a given axial force N_u, there is a sole value of M_u, and vice versa. The symmetrical reinforcement section is used as an example to establish the correlation curve equation of N_u-M_u.

5.10.1 N_u-M_u Interaction Diagram of Rectangular Sectional Large Eccentrically Compressed Members with Symmetric Reinforcement

Plug N_u, $A_s = A_s'$, $f_y = f_y'$ into Eq. (5-13)

$$N_u = \alpha_1 f_c b x \tag{5-53}$$

$$x = \frac{N_u}{\alpha_1 f_c b} \tag{5-54}$$

Integrating Eqs. (5-54) and (5-15) into (5-14)

$$N_u \left(e_i + \frac{h}{2} - a_s \right) = \alpha_1 f_c b \frac{N_u}{\alpha_1 f_c b} \left(h_0 - \frac{N_u}{2\alpha_1 f_c b} \right) + f_y' A_s' (h_0 - a_s') \tag{5-55}$$

Thus

$$N_u e_i = -\frac{N_u^2}{2\alpha_1 f_c b} + \frac{N_u h}{2} + f_y' A_s' (h_0 - a_s') \tag{5-56}$$

Due to $N_u e_i = M_u$

$$N_u = -\frac{N_u^2}{2\alpha_1 f_c b} + \frac{N_u h}{2} + f_y' A_s' (h_0 - a_s') \tag{5-57}$$

This is the N_u-M_u correlation curve of large eccentric compression rectangular section with symmetric reinforcement.

It can be seen from Eq. (5-57) that M_u is a quadratic function of N_u. M_u increases with the increasing of N_u, as shown in the curves below the horizontal dashed line in Figure 5-28.

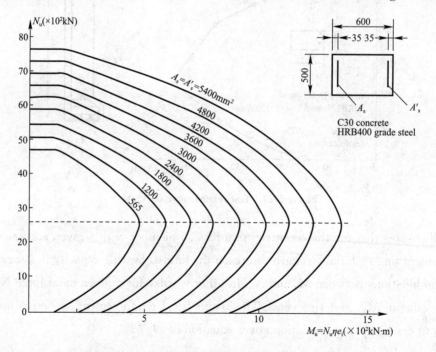

Figure 5-28 N_u-M_u interaction curve of symmetrical reinforcement

5.10.2 N_u-M_u Interaction Diagram of Rectangular Sectional Small Eccentrically Compressed Members with Symmetric Reinforcement

Assuming that the section is partially compressed, N_u, σ_s and $x = \xi h_0$ are taken into Eq. (5-20), and N_u, $x = \xi h_0$ are taken into Eq. (5-21), which can be obtained

$$N_u = \alpha_1 f_c b h_0 \xi + f'_y A'_s - \left(\frac{\xi - \beta_1}{\xi_b - \beta_1}\right) f_y A_s \tag{5-58}$$

$$N_u e = \alpha_1 f_c b h_0^2 \xi (1 - 0.5\xi) + f'_y A'_s (h_0 - a'_s) \tag{5-59}$$

Taking $A_s = A'_s$, $f_y = f'_y$ into Eq. (5-61), N_u can be obtained as:

$$N_u = \frac{\alpha_1 f_c b h_0 (\xi_b - \beta_1) - f'_y A'_s}{\xi_b - \beta_1} \xi - \left(\frac{\xi_b}{\xi_b - \beta_1}\right) f_y A'_s$$

The following is acquired by the above equation:

$$\xi = \frac{\beta_1 - \xi_b}{\alpha_1 f_c b h_0 (\beta_1 - \xi_b) + f'_y A'_s} N_u - \frac{\xi_b f'_y A'_s}{\alpha_1 f_c b h_0 (\beta_1 - \xi_b) + f'_y A'_s} \tag{5-60}$$

Taking
$$\lambda_1 = \frac{\beta_1 - \xi_b}{\alpha_1 f_c b h_0 (\beta_1 - \xi_b) + f'_y A'_s} \tag{5-61}$$

$$\lambda_2 = \frac{\xi_b f'_y A'_s}{\alpha_1 f_c b h_0 (\beta_1 - \xi_b) + f'_y A'_s} \tag{5-62}$$

Then, $\xi = \lambda_1 N_u + \lambda_2$

Taking Eqs. (5-61), (5-62) and (5-15) into Eq. (5-59)

$$N_u \left(e_i + \frac{h}{2} - a_s\right) = \alpha_1 f_c b h_0^2 (\lambda_1 N_u + \lambda_2)\left(1 - \frac{\lambda_1 N_u + \lambda_2}{2}\right) + f'_y A'_s (h_0 - a'_s)$$

M_u can be obtained according to $N_u e_i = M_u$.

$$M_u = \alpha_1 f_c b h_0^2 [(\lambda_1 N_u + \lambda_2) - 0.5 (\lambda_1 N_u + \lambda_2)^2] - \left(\frac{h}{2} - a_s\right) N_u + f'_y A'_s (h_0 - a'_s) \tag{5-63}$$

This is the relevant equation of N_u-M_u of rectangular sectional small eccentrically compressed member with symmetric reinforcement. M_u is a quadratic function of N_u, but M_u will decrease with the increasing of N_u, as shown in the curves above the dashed line in Figure 5-58.

5.10.3 Features and Applications of N_u-M_u Interaction Diagram

The N_u-M_u interaction diagram consists of two curve segments: large eccentrically compressed failure and small eccentrically compressed failure. Its characteristics are:

(1) When $M_u = 0$, N_u is the largest; when $N_u = 0$, M_u is not the largest; when the boundary failure occurs, M_u is the largest;

(2) When it is small eccentric compression, N_u decreases as M_u increases. However in large eccentric compression, N_u increases as M_u increases;

(3) When members are in symmetrical reinforcement under the same cross-sectional shape, size, the concrete grade and steel grade but with different number of reinforcement, N_u are the same under boundary failures (due to $N_u=\alpha_1 f_c b x_b$). Therefore, the breakdown points of each N_u-M_u curve are at the same horizontal level, as the dashed line shown in Figure 5-28.

Taking advantage of related equations, a series of charts can be plotted in advance for some eccentric compressed members with specific section sizes, concrete strength grades and steel categories. Required reinforcement area can be obtained directly by looking up relevant charts to simplify the calculation, which will save a lot of work. The normal sectional bearing capacity chart of rectangular sectional eccentrically compressed member in symmetrical reinforcement with section size of $b \times h = 500\text{mm} \times 600\text{mm}$, concrete strength grade of C30 and steel bar of HRB400 is shown in Figure 5-28. At the beginning of the design, e_i is calculated at first, followed by looking charts up adapted to design conditions. A_s and A'_s can also be looked up by the values of N and Ne_i.

5.11 Shear Capacity Formulae for Eccentrically Compressed Members

The shear value of the eccentrically flexural member is relatively small in general, with no need to calculate shear capacity of diagonal section. However, for the frame columns under relatively large horizontal forces and truss chords under transverse load, shear force has relatively great impact, which has to be considered.

Experimental results show that the axial compressure can delay vertical cracks' appearing and reduce the crack widths. Under the axial compression, it will appear the situation of compression height increasing, slope angle of diagonal cracks being small, the horizontal projection length basically unchanging and tension of longitudinal bars decreasing, which increase the shear capacity of diagonal section. But the positive effect has a limitation when axial compression ratio reaches $N/f_c bh = 0.3 \sim 0.5$, the small eccentricity compressed failure with diagonal cracks will occur as increasing the axial compression, leading to the maximum of shear capacity of diagonal section, as shown in Figure 5-29.

Besides, the experiment shows that when $N < 0.3 f_c bh$, the influence of axial compression among members with different shear span ratios is tiny, as shown in Figure 5-30.

Based on analysis of test data and reliability calculation, for eccentrically compressed members of rectangular, T-shaped and I-shaped sections under axial force and lateral force, the shear capacity of inclined section is recommended to be calculated by the following equation:

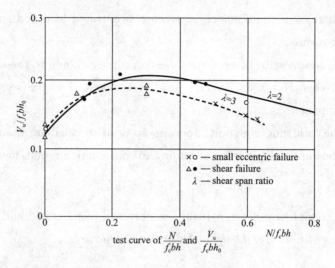

Figure 5-29 Relation between relative axial compression and shear force

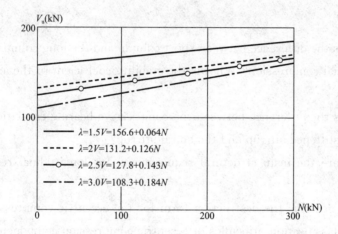

Figure 5-30 Comparison diagram of regression equation of V_u and N under different shear span ratios

$$V_u = \frac{1.75}{\lambda + 1.0} f_t b h_0 + 1.0 f_{yv} \frac{A_{sv}}{s} h_0 + 0.07 N \qquad (5\text{-}64)$$

Where λ——shear span ratio of calculated section of eccentrically compressed members. For all kinds of frame columns, $\lambda = M/Vh_0$ can be taken. When the point of inflection for the frame column is within the scope of the height, $\lambda = H_n/2h_0$ (H_n is the clear height of column) can be taken; when $\lambda < 1$, $\lambda = 1$ is taken; when $\lambda > 3$, $\lambda = 3$ is taken; M is the moment design value corresponding to the shear design value V on the calculated section. For other eccentrically compressed members, when bearing the uniformly distributed load, $\lambda = 1.5$ is taken; when bearing the concentrated load taking up more than 75% of the total shear values in the support section or at the edge of the joint, $\lambda = a/h_0$ is taken; when $\lambda < 1.5$, $\lambda = 1.5$ is taken; when $\lambda > 3$, $\lambda = 3$ is taken.

Where, a is the distance between concentrated load and support or edge of the member.

N——the design value of axial compression corresponding to the shear design value V. When $N > 0.3 f_c A$, $N = 0.3 f_c A$ is taken. A is the sectional area of the component.

When met the following equation, there is no need to calculate the shear capacity of the inclined section, and only stirrups according to construction requirements are arranged.

$$V \leqslant \frac{1.75}{\lambda + 1.0} f_t b h_0 + 0.07 N \tag{5-65}$$

The shear sectional size of eccentrically compressed members should comply with the relevant provisions of *Code for Design of Concrete Structures*.

Questions

5-1 What is the difference between short column and the long column of the ordinary stirrups under axial compression? What is the stability coefficient of the axial compression column?

5-2 What is the difference between the compressive bearing capacities of the normal section with the stiffened stirrup and the spiral stirrup?

5-3 What are the main structural requirements for longitudinal reinforcement and stirrups?

5-4 Briefly describe the destruction form for the eccentric compression of the short column and the classification principles of eccentric compression components?

5-5 What is the difference between the failure modes of the normal section for the long column and for the short column? What is the second-order effect of the P-δ for eccentric compression members?

5-6 Under what circumstances should the P-δ effect be considered?

5-7 How to distinguish between large and small eccentric compression damages?

5-8 What is the simplified calculation for the compressive bearing capacity of the normal section with the asymmetric reinforcement in the large-eccentric compression member?

5-9 How to calculate the compression capacity of the rectangular section with asymmetric reinforcement for small eccentric compression members?

5-10 How to design the asymmetric reinforcement in the rectangular cross-section of large eccentric compression members and small eccentric compression members by the compression capacity of the cross-section design?

5-11 How to distinguish the boundaries between the large and small eccentric compression damages in the rectangular section with symmetrical reinforcement?

5-12 How to design a rectangular cross-section symmetrical reinforcement of large eccentric compression members and small eccentric compression members by the normal cross-section bearing capacity design?

5-13 What is the correlation curve N_u-M_u for the normal cross-section bearing capacity in the eccentric compression member?

5-14 How to calculate the shear capacity of the oblique section for the eccentric compression member?

Exercises

5-1 For a second layer of four-span cast-in-situ frame structure, the design value of axial pressure is $N=1100$kN with the height of 6m. Strength grade of concrete is C30. Steel bars are HRB335. The sectional size is $b \times h = 350\text{mm} \times 350\text{mm}$. A_s is required.

5-2 An eccentrically compressed column is a rectangular section. The dimensions of the section are $b \times h = 400\text{mm} \times 600\text{mm}$, $a_s = a_s' = 400\text{mm}$, and $l_0 = 6\text{m}$. The strength of the concrete is $f_c = 16.7\text{N/mm}^2$. The properties of the reinforcement are $f_y' = f_y = 300\text{N/mm}^2$, $E_s = 2 \times 10^5 \text{N/mm}^2$. If the axial compressive force $N_c = 1200\text{kN}$ and the moment $M = 600\text{kN} \cdot \text{m}$ are applied on the column and the section is asymmetrically reinforced,

(1) Determine A_s and A_s'.

(2) If 4⚏20 compression steel bars have already been arranged($A_s' = 1257\text{mm}^2$), A_s is to be required.

(3) Compare the calculation results in question (1) and (2) and analyze the difference.

5-3 An eccentrically compressed column is a rectangular section. The dimensions of the section are $b \times h = 500\text{mm} \times 800\text{mm}$, $a_s = a_s' = 40\text{mm}$ and $l_0 = 12.5\text{m}$. The strength of the concrete is $f_c = 14.3\text{N/mm}^2$. The properties of the reinforcement are $f_y' = f_y = 300\text{N/mm}^2$, $E_s = 2 \times 10^5 \text{N/mm}^2$. If the axial compressive force $N_c = 1800\text{kN}$ and the moment $M = 1080\text{kN} \cdot \text{m}$ are applied on the member and the section is asymmetrically reinforced, A_s and A_s' are to be determined.

5-4 An eccentrically compressed column is a rectangular section. The dimensions of the section are $b \times h = 500\text{mm} \times 800\text{mm}$, $a_s = a_s' = 45\text{mm}$ and $l_0 = 4.6\text{m}$. The strength of the concrete is $f_c = 14.3\text{N/mm}^2$. The properties of the reinforcement are $f_y' = f_y = 300\text{N/mm}^2$, $E_s = 2 \times 10^5 \text{N/mm}^2$. If the axial compressive force $N_c = 7000\text{kN}$ and the moment $M = 175\text{kN} \cdot \text{m}$ are applied on the column and the section is asymmetrically reinforced, A_s and A_s' are to be determined.

5-5 In Exercise 5-3, if the section is symmetrically reinforced, determine A_s and A_s'.

5-6 In Exercise 5-4, if the section is symmetrically reinforced, determine A_s and A_s'.

5-7 An I-section column in a structural laboratory has the dimensions of $b=1200$mm, $h=800$mm, $b_f'=b_f=400$mm, $h_f'=h_f=130$mm, $a_s=a_s'=40$mm and $l_0=6.8$m. The strength of the concrete is $f_c=14.3$N/mm². The properties of the reinforcement are $f_y'=f_y=300$N/mm². If the axial force $N_c=1000$kN and the moment $M=400$kN·m are applied on the column and the section is asymmetrically reinforced, A_s and A_s' are to be determined.

Chapter 6
Load-Carrying Capacity for Tension Members

6.1 Calculation for Cross Section Tension Capacity of Axial Tension Members

Figure 6-1 illustrates an axial tension test on a reinforced concrete member. Prior to concrete cracking, the tensile force N_t is carried by both the concrete and the steel bar. After cracking, the concrete at the cracks cannot take any load and the tensile force N_t is completely carried by the steel bar. When the steel yields, the specimen reaches its ultimate tension capacity. The whole process of the tensile test can be divided into three stages.

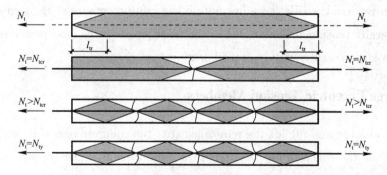

Figure 6-1　Cracks of an axially tensioned reinforced concrete member

Similar with the appropriate reinforcement beam, axially tensioned structural members can be divided into three stages. Stage I starts from the onset of loading and ends just before the concrete cracks. Stage II is from the concrete crack to the yielding of the steel bar. When tensile force reaches the yield load, the steel bar begins to yield and Stage III begins. When all the steel bars have yielded, the specimen is to fail. Stage III is used to calculate the tensile capacities of members.

When the reinforcement yields, the tensioned member will undergo a larger deformation further while the tensile load holds constant. Therefore this load is taken as the ultimate tensile capacity of the member and can be expressed as

$$N_u = f_y A_s \qquad (6\text{-}1)$$

[**Example 6-1**] A structural member with the cross-sectional dimension of $b \times h = 300\text{mm} \times 300\text{mm}$ is subjected to an axial tension of 680kN. The yield strength of the steel is 360N/mm^2. The amount of the reinforcement is required based on the ultimate tensile capacity.

[**Solution**]

$$N = 680\text{kN}, \quad f_y = 360\text{N/mm}^2$$

Thus,

$$A_s = \frac{N}{f_y} = \frac{680000}{360} = 1889\text{mm}^2$$

6 Φ 20 steel bars are chosen with area of $A_s = 1884\text{mm}^2$.

6.2 Calculation for Cross Section Tension Capacity of Eccentric Tension Members

The section of a reinforced concrete eccentrically tensioned member may be divided into two types, i.e., small eccentrically tensioned section and large eccentrically tensioned section. When the axial tensile force lies within the reinforcements at the two sides, it is a small eccentrically tensioned section. When the axial tensile force lies outside the reinforcements, it is a large eccentrically tensioned section.

6.2.1 Large Eccentric Tension Members

If the axial force lies outside the reinforcement, the concrete near the axial force is under tension while the concrete far away from the axial force is under compression. The tensile stress near the axial force increases with the axial force and cracks until reaching the tensile strength. However, there is always a compression zone in the section. Otherwise, the section cannot maintain the force equilibrium.

When the axial force is big enough to cause the reinforcement yielding near the axial force, the propagation of the crack will further reduce the compression zone, which leads to the increasing of the compressive stress in the concrete. Finally, the concrete is crushed when the strain in the extreme compression fiber reaches ε_{cu}. It can be observed that this failure mode is similar to that of large eccentricity compression sections. However, if too many steel bars are arranged near the axial force and few steel bars are far away from the axial force, it is likely that the concrete far away from the axial force is crushed at first, while the reinforcement near the axial force has not yielded, similar to that of an over-reinforced flexural section.

Based on Figure 6-2 that shows the simplified stress distribution across the large eccentrically tensioned section, the equilibrium equations can be obtained as

$$N_u = f_y A_s - f'_y A'_s - \alpha_1 f_c bx \tag{6-2}$$

$$N_u e = \alpha_1 f_c hx \left(h_0 - \frac{x}{2}\right) + f'_y A'_s (h_0 - a'_s) \tag{6-3}$$

Where e is the distance from the axial force to the resultant force of A_s.

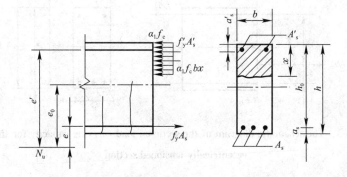

Figure 6-2 The schematic diagram of the sectional load-carrying capacity for the large eccentrically tensioned section

$$e = e_0 - \frac{h}{2} + a_s \tag{6-4}$$

When $x < 2a'_s$, $x = 2a'_s$ is assumed to ensure the yielding of A'_s.

To minimize the steel quantity, the same method in the design of large eccentrically compressive sections can be used, i.e., $x = x_b = \xi_b h_0$. Then, A'_s and A_s can be determined by Eq. (6-5) and by Eq. (6-6), respectively.

$$A'_s = \frac{N_u e - \alpha_1 f_c b x_b (h_0 - x_b/2)}{f'_y (h_0 - a'_s)} \tag{6-5}$$

$$A_s = \frac{\alpha_1 f_c b x_b + N_u}{f_y} + \frac{f'_y}{f_y} A'_s \tag{6-6}$$

6.2.2 Small Eccentric Tension Members

Small eccentrically tensioned members can be classified as fully tensioned members and partially tensioned members, which depends on the value of e_0. For fully tensioned sections, e_0 is small. So the concrete across the full section is subjected to tension, and the tensile stress in the concrete near the axial force is larger. For partially tensioned sections, e_0 is large and the concrete far from the axial force is under compression.

With the increase of the axial force, the stress in the concrete also increases. When the ultimate tensile stress of fiber reaches the tensile strength of concrete, the section is cracked. For the smaller e_0, the crack will rapidly run through the section transversely. For the larger e_0, the compressive stress in the original compression zone will change into tensile stress for equilibrium, because the concrete in the cracked tension zone is out of work.

Therefore, for a small eccentricity section in the ultimate state, the concrete across the full section cracks and all the longitudinal bars can yield, as shown in Figure 6-3.

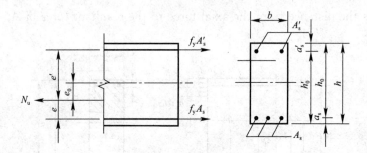

Figure 6-3 The schematic diagram of the sectional load-carrying capacity for the small eccentrically tensioned section

From Figure 6-3, the equilibrium equations can be obtained as

$$N_u = f_y A_s + f'_y A'_s \tag{6-7}$$

$$N_u e = f_y A'_s (h_0 - a'_s) \tag{6-8}$$

$$N_u e' = f_y A_s (h'_0 - a_s) \tag{6-9}$$

$$e = \frac{h}{2} - e_0 - a_s \tag{6-10}$$

$$e' = e_0 + \frac{h}{2} - a'_s \tag{6-11}$$

Where e —— The distance from the axial force to the resultant force of A_s;

e' —— The distance from the axial force to the resultant force of A'_s.

For symmetric section

$$A'_s = A_s = \frac{N_u e'}{f_y (h_0 - a'_s)} \tag{6-12}$$

$$e' = e_0 + \frac{h}{2} - a'_s \tag{6-13}$$

[**Example 6-2**] Given a rectangular section of $b = 350$mm and $h = 400$mm, the concrete is of grade C25 and the reinforced steels are of HRB335. The reinforcements are to be designed for an axial tension $N = 500$kN and a bending moment $M = 40$kN·m.

[**Data**] concrete C25: $f_c = 11.9$N/mm², $f_t = 1.27$N/mm²

Steel HRB335: $f_y = f'_y = 300$N/mm², $\xi_b = 0.55$

$h_0 = h - a_s = 400 - 35 = 365$mm

Minimum steel ratio: $\rho_{min} = \max \begin{cases} 0.2\% \\ 0.45 \dfrac{f_t}{f_y} = 0.2\% \end{cases}$

[Solution]

(1) It is to be determined whether it is a large eccentrically tensioned section or not

$$e_0 = \frac{M}{N} = \frac{40}{500} = 0.08\text{m} = 80\text{mm} < \frac{h}{2} - a_s = \frac{400}{2} - 35 = 165\text{mm}$$

The axial force is applied between A'_s and A_s, and it is a small eccentrically tensioned section.

(2) Calculation of A'_s and A_s

$$e = \frac{h}{2} - e_0 - a_s = \frac{400}{2} - 80 - 35 = 85\text{mm}$$

$$e' = \frac{h}{2} + e_0 - a'_s = \frac{400}{2} + 80 - 35 = 245\text{mm}$$

$$A_s = \frac{Ne'}{f_y(h'_0 - a_s)} = \frac{500 \times 10^3 \times 245}{300(365 - 35)} = 1237.4\text{mm}^2$$

$$A'_s = \frac{Ne}{f_y(h_0 - a'_s)} = \frac{500 \times 10^3 \times 85}{300(365 - 35)} = 429.3\text{mm}^2$$

$$A_s > A'_s > \rho_{min}bh = 0.2\% \times 350 \times 400 = 280\text{mm}^2$$

4 ⌽ 20 is adopted with area of 1256mm² for A_s, and 4 ⌽ 12 is adopted with area of 452mm² for A'_s.

[Example 6-3] The thickness of the wall for a water tank is 400mm with $a_s = a'_s = 40$mm. The concrete strength is $f_c = 14.3\text{N/mm}^2$ and $f_t = 1.43\text{N/mm}^2$. The strength and modulus of elasticity of the steels are $f_y = f'_y = 300\text{N/mm}^2$ and $E_s = 2 \times 10^5 \text{N/mm}^2$. The maximum moment per unit width (in meter) in the horizontal direction at the mid-span of the wall is $M = 320\text{kN} \cdot \text{m}$ and the corresponding tensile force is $N = 400\text{kN}$, A'_s and A_s are to be determined.

[Solution]

(1) It is to be determined whether it is a large eccentrically tensioned section or not

$$e_0 = \frac{M}{N} = \frac{320}{400} = 0.8\text{m} = 800\text{mm} > \frac{h}{2} - a_s = \frac{400}{2} - 40 = 160\text{mm}$$

It is a large eccentrically tensioned section.

(2) Calculate A'_s and A_s

$$e = e_0 - \frac{h}{2} + a_s = 800 - \frac{400}{2} + 40 = 640\text{mm}$$

Taking $\xi = \xi_b = 0.55$, the value for A'_s is:

$$A'_s = \frac{N_u e - \alpha_1 f_c b x_b (h_0 - x_b/2)}{f_y(h_0 - a'_s)}$$

$$= \frac{400 \times 10^3 \times 640 - 14.3 \times 1000 \times 10^3 \times (0.55 - 0.5 \times 0.55^2)}{300 \times (360 - 40)} = -5031 < 0$$

Thus, $A'_s = \rho'_{min} bh = 0.2\% \times 1000 \times 400 = 800 \text{mm}^2$ and $\Phi 16@250$ is adopted ($A'_s = 804 \text{mm}^2$).

Next step is to determine A_s with the known A'_s:

$$N_u e = \alpha_1 f_c bx \left(h_0 - \frac{x}{2}\right) + f'_y A'_s (h_0 - a'_s)$$

$400 \times 10^3 \times 640 = 14.3 \times 1000 \times 360^2 \times (\xi - 0.5\xi^2) + 300 \times 804 \times (360 - 40)$

$\xi^2 - 2\xi + 0.193 = 0$

The result for the above equation is $\xi = 0.102$, $x = \xi h_0 = 36.7 \text{mm} < 80 \text{mm}$, $x = 80 \text{mm}$ is taken.

$$e' = \frac{h}{2} + e_0 - a'_s = \frac{400}{2} + 800 - 40 = 960 \text{mm}$$

$$A_s = \frac{N_u e'}{f_y (h'_0 - a_s)} = \frac{400 \times 10^3 \times 960}{300 \times (360 - 40)} = 4000 \text{mm}^2$$

$A_s > 0.45 \frac{f_t}{f_y} bh = 0.45 \times \frac{1.43}{300} \times 1000 \times 400 = 858 \text{mm}^2$, the requirement for minimum ratio is satisfied.

6.3 Calculation for Shear Capacity of Eccentric Tension Members

The axial tension increases the principal tensile stresses in the member. The angle between the inclined crack and the longitudinal axis of the member is also enlarged, which will reduce the depth of the shear-compression zone at failure. When the axial tension is large, no shear-compression zone will exist. In a word, the axial tension exerts negative effects to the shear capacity of the member, and the shear capacity will decrease larger with the increase of the axial tension.

Therefore, the shear capacities of eccentrically tensioned members of rectangular, T-sections and I-sections can be calculated by

$$V_u = \frac{1.75}{\lambda + 1.0} f_t bh_0 + f_{yv} \frac{A_{sv}}{s} h_0 - 0.2N \qquad (6-14)$$

Where λ——Shear-span ratio of calculated sections, determined in the same way as the eccentrically compressed members;

N——Axial tension on members.

In Eq. (6-14), V_u should not be less than $f_{yv} \frac{A_{sv}}{s} h_0$ and $f_{yv} \frac{A_{sv}}{s} h_0$ should not be less than

$0.36 f_t b h_0$.

Questions

6-1 Why won't new cracks appear in an axially tensioned concrete member when the number of cracks has reached a certain value?

6-2 How is the cracking load and the ultimate load of an axially tensioned structural member determined?

Exercises

6-1 The thickness of the wall of a water tank is 500mm with $a_s = a'_s = 40$mm. The concrete strength is $f_c = 14.3$N/mm² and $f_t = 1.43$N/mm². The strength and modulus of elasticity of the steels are $f_y = f'_y = 300$N/mm² and $E_s = 2 \times 10^5$N/mm² respectively. If the maximum moment per unit width (meter) in the horizontal direction at the mid-span of the wall is $M = 380$kN·m and the corresponding tensile force is $N = 450$kN, A'_s and A_s are required.

Chapter 7
Load-Carrying Capacity of Torsional Members

7.1 Introduction

Torsion is one of the basic loading types of structural members. In engineering structures, torsion seldom occurs alone. Generally, the members are under combination of torsion, shear and flexure. This is illustrated by the typical torsional structure shown in Fig 7-1(a) and (b). These members are under the combined action of flexure, shear and torsion.

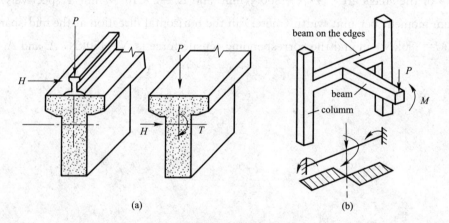

Figure 7-1 Examples of structural elements under torsion
(a) Crane Beam; (b) Beam on the Edges

For a statically determinate structure subjected to torsional loading, the torque is maintained by the static equilibrium conditions. And the torsional stresses may not be reduced by redistribution, which is referred to as equilibrium torsion. The T in cross sections is called equilibrium torque. The torsional reaction T should be able to maintain equilibrium and the section has to be designed with this capacity, otherwise collapse may occur, as shown in Figure 7-1(a).

For an indeterminate structure, the torque acting on the members is an addition to static equilibrium conditions, which means compatibility of deformation with the adjacent members should be maintained. The spandrel beam shown in Figure 7-1(b) is subjected to torsional force from the supported slab. When the beam on the edges and the floor beam crack, the torsional stiffness of the side beams significantly changes due to the reduction on the bending stiffness of the floor beams. Floor beams and side beams will generate internal force redistribution.

7.2 Experimental Study on Members Subjected to Pure Torsion

7.2.1 Performance Prior to Cracking

Before cracking, for the pure torsion, it is found from the recorded torque-twist angle curve (Figure 7-2) that the behavior of the members conforms to the Saint Venant's Principle prior to cracking. Hence, when the torque is low, the curve is almost a straight line. The stresses of longitudinal bars and stirrups are very small. When the torque increases to the cracking torque T_{cr}, the curve deviates from the initial direction.

7.2.2 Performance after Cracking

When the cracks appear, because the concrete quit work, the reinforcement stress increases obviously, in particular the angle begins to increase dramatically. At this time, the previous equilibrium is not matched, and a new resistance mechanism, which combines concrete of cracks and reinforcement is formed in the member after cracking.

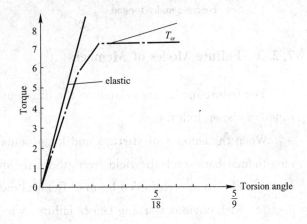

Figure 7-2 The performance prior to cracking

After cracking, the members can still bear the torque; however, the torsional rigidity of the members drops dramatically, which is indicated in Figure 7-3 by an apparent slope change, and the reduction in the reinforcement will lead to the torsional rigidities reducing more. Experiments show that the concrete is in compression and the longitudinal bars and the stirrups are in tension.

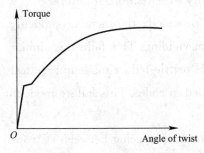

Figure 7-3 The curve of T-θ

The T_{cr} is higher about 10%~30% than in plain concrete members.

According to the experiments, the initial crack happens at approximately midpoint of the longer dimension of the cross section. The inclination of the crack is 45° to the longitudinal axis of the member. Under loading, this crack propagates in a spiral way and maintains its 45° angle to the longitudinal axis in Figure 7-4. As the torque continues to increase, the mechanism remains unchanged and the stresses in the concrete and reinforcement increase continuously until the final failure (Figure 7-5).

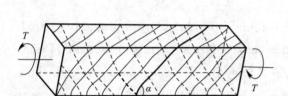
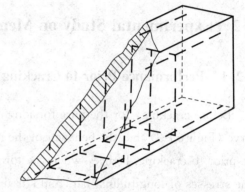

Figure 7-4 The spiral crack of reinforced concrete under torsion

Figure 7-5 Spatial distorting surface of reinforced concrete members with pure torsion

7.2.3 Failure Modes of Members

The failure modes are related to the different quantities of longitudinal bars and stirrups, as shown follows:

When the amount of stirrups and longitudinal bars are appropriate, the stirrups and longitudinal bars reach the yield strength firstly and then the concrete is crushed. This failure is similar to the reinforced beam of flexural members. The total process exhibits certain ductility with obvious warning before failure, which is a dominant failure mode that is permitted in practical design. This failure mode is termed as the bending failure mode.

If either longitudinal bars or stirrups are overly provided, the overly provided reinforcement will not yield at failure. This failure is similar to the partially over-reinforced members on torsion which exhibits ductility to some extent, but not as much as that of the under-reinforced failure. To avoid this failure mode, the ratio of longitudinal bars to stirrups is specified. This failure mode is termed as the partially over-reinforced failure.

When stirrups and longitudinal bars are both overly provided, the member suddenly fails due to the crushing of concrete before neither of them yielding. This failure is similar to over reinforced beam on flexural members. To avoid this brittle failure mode in practical design, the minimum cross-sectional area is usually specified in codes. This failure mode is termed as the compression failure.

When stirrups and longitudinal bars are not sufficient or the spacing between stirrups is too large, once the member cracks, the reinforcement cannot resist the tensile stress transferred from the concrete, so the member fails immediately. This failure is similar to lightly reinforced beam on flexural members. The failure is brittle and appears with little warning, which is termed as the tension failure.

Therefore, we should avoid the tension failure and compression failure in design.

7.3 Torsional Capacities of the Cross-section in Pure Torsion

7.3.1 Cracking Torque

As previously mentioned, the stress of the reinforcement is very low with little effect on the cracking torque, so it is usually neglected in the calculation before cracking.

As shown in Figure 7-6, the torque generates the shear stress τ on the cross section. The directions of the principal tensile stress σ_{tp} and the principal compressive stress σ_{cp} are $45°$ and $135°$ respectively to the longitudinal axis.

$$|\sigma_{tp}|=|\sigma_{cp}|=\tau$$

For the ideal elastic-plastic materials, the distribution of shear stresses in a torsional member of rectangular cross section is shown in Figure 7-7(a). The shear stresses at the section's corners are zero

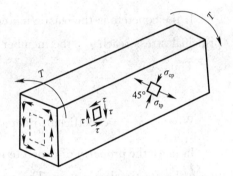

Figure 7-6 The torsion of rectangular cross sections

and the maximum shear stress τ_{max} and the maximum principal stress appear at the center of the longer side of the cross section. When the maximum shear stress and the maximum principal stress reach the concrete tensile strength, the loads can increase a little until the tensile strain of cross-sectional edge reaches the concrete ultimate tensile strain, and finally the members crack. Meantime, the torque is termed as cracking torque T_{cr}, as is shown in Figure 7-7(b).

According to the plastic theory, dividing the rectangular section into four portions, Figure 7-7(c) shows how the resultant forces of each portion form two pairs of couples which are cracking torsions of the section as shown in Eq. (7-1).

$$T_{cr} = \tau_{max} W_t = \tau_{max} \frac{b^2}{6}(3h-b) = f_t \frac{b^2}{6}(3h-b) \qquad (7\text{-}1)$$

Where h, b——the dimensions of the cross section, $h > b$;

W_t——the plastic torsional modulus of the cross section for rectangular section.

$$W_t = \frac{b^2}{6}(3h-b)$$

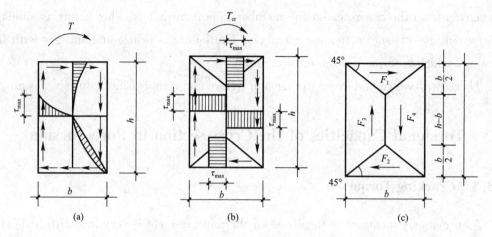

Figure 7-7 The stress distribution of maximum shear

If the concrete is the elastic material when the maximum shear stress or the maximum principal stress reach f_t, the member cracks.

Thus:

$$T_{cr} = f_t \alpha b^2 h \tag{7-2}$$

Where α——a coefficient related to the h/b, $\dfrac{h}{b}=1\sim10$, $\alpha=0.208\sim0.313$.

In fact, the property of concrete is neither elastic material nor plastic material, but between the elastic-plastic state. The research shows that calculated cracking torque is lower than the Eq. (7-1) and is higher than the Eq. (7-2).

For convenience, concrete is not an ideal elastic material, the cracking torque should be appropriately discounted. Experimental results show that the reduction factor is 0.7 for high strength concrete and 0.8 for low strength concrete. So for safety, the cracking torque is taken asfollows:

$$T_{cr} = 0.7 f_t W_t \tag{7-3}$$

7.3.2 Calculation for Torsional Capacities of Members by Variable Angle Space Truss Subjected to Pure Torsion

The research shows that, in pure torsion, once the inclined cracks in plain concrete members appear, the members will destroy. If the arrangements of stirrups and longitudinal bars are appropriate, the member will not immediately fail after the occurrence of initial cracks and the total process exhibits certain ductility.

So far, the torsional capacities of twisted sections in reinforced concrete members have been calculated by two quite different theories—the variable angle space truss model analogy and the skew bending theory. And *Code for Design of Concrete Structures* adopts the former method.

In 1968, P. Lampert and B. Thuerlimann came up with the variable angle space truss model (Figure 7-8). It is the improvement and development of the space truss model proposed by E. Rausch in 1929. This model overcame the above-mentioned shortcomings and will be detailed herein.

The basic idea of variable angle space truss model is: when the cracks have fully developed and reinforced stress is close to the yield strength, the sectional core concrete quit working, so the reinforced concrete member of solid section in torsion can be regarded as a hollow box-sectional member, which consists of spiral cracks of concrete shell, longitudinal reinforcement and stirrups together as the variable angle space truss to resist torque.

The variable angle space truss model obeys the following assumptions:

(1) The original solid cross section is simplified as a tubular section. The concrete in the tubular section is split by spiral cracks, which leads to many compression struts with the inclination of α;

(2) The longitudinal bars and stirrups serve as tensile members and form the space truss with the compression struts;

(3) The dowel action of the bars is ignored.

Based on the thin-walled tube theory, the shear flow q (Figure 7-8b) induced by torque T in the tubular section can be expressed as follows:

$$q = \tau t_d = \frac{T}{2A_{cor}} \tag{7-4}$$

Where A_{cor}——the area taken enclosed by the center line of the tubular section wall. It is herein the area enclosed by the connection line of centroids of longitudinal bars at corners of the section, $A_{cor} = b_{cor} \times h_{cor}$.

τ——the shear stress due to torsion.

t_d——the thickness of the wall in the tubular section.

As is shown in Figure 7-8(a), the variable angle space truss model consists of a double variable angle space vertical truss and a double variable angle space horizontal truss. Firstly, the variable angle space vertical truss is studied.

Figure 7-8(c) also shows the shear flow q induced by forces in the truss members. For the diagonal struts, the average compressive stress is σ_c and the resultant of the compressive force is D. According to the static equilibrium conditions:

Inclined compression

$$D = \frac{qb_{cor}}{\sin\alpha} = \frac{\tau t_d b_{cor}}{\sin\alpha} \tag{7-5}$$

The average compressive stress of concrete

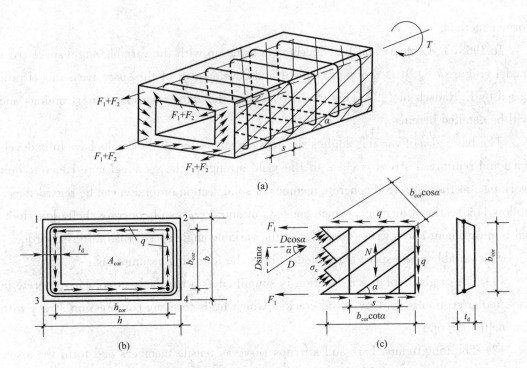

Figure 7-8 Variable angle space truss model

$$\sigma_c = \frac{D}{t_d b_{cor} \cos\alpha} = \frac{q}{t_d \sin\alpha \cos\alpha} = \frac{\tau}{\sin\alpha \cos\alpha} \quad (7\text{-}6)$$

The tension of longitudinal reinforcement

$$F_1 = \frac{1}{2}D\cos\alpha = \frac{1}{2}qb_{cor}\cot\alpha = \frac{1}{2}\tau t_d b_{cor}\cot\alpha = \frac{Tb_{cor}}{2A_{cor}}\cot\alpha \quad (7\text{-}7)$$

The tension of stirrups

$$N = \frac{qb_{cor}}{b_{cor}\cot\alpha}s$$

Thus

$$N = qs\tan\alpha = \tau t_d s \tan\alpha = \frac{T}{2A_{cor}}s\tan\alpha \quad (7\text{-}8)$$

It is supposed that the diagonal strut angle of variable angle space level truss is α, Thus

$$F_2 = \frac{Th_{cor}}{2A_{cor}}\cot\alpha \quad (7\text{-}9)$$

Therefore, the total of all the tensions of longitudinal bars is

$$R = 4(F_1 + F_2) = q\cot\alpha \, u_{cor} = \frac{Tu_{cor}}{2A_{cor}}\cot\alpha \quad (7\text{-}10)$$

Where u_{cor}——the perimeter corresponding to the area A_{cor} enclosed by the shear flow,

$u_{cor} = 2(b_{cor} + h_{cor})$

The average compressive stress of concrete

$$\sigma_c = \frac{T}{2A_{cor}t_d \sin\alpha\cos\alpha} \tag{7-11}$$

Eqs. (7-4), (7-8), (7-10) and (7-11) are static equilibrium equations according to variable angle space truss. If the amount of the longitudinal bars is appropriate, the bars yield prior to failure. Thus, the stress can be taken as f_y and f_{yv}, separately. Then,

$$R = R_y = f_y A_{stl} \tag{7-12}$$

$$N = N_y = f_{yv} A_{stl} \tag{7-13}$$

According to Eqs. (7-10) and (7-8), the torsional capacity of members is

$$T_u = 2R_y \frac{A_{cor}}{u_{cor}} \tan\alpha = 2f_y A_{stl} \frac{A_{cor}}{u_{cor}} \tan\alpha \tag{7-14}$$

$$T_u = 2N_y \frac{A_{cor}}{s} \cot\alpha = 2f_{yv} A_{stl} \frac{A_{cor}}{s} \cot\alpha \tag{7-15}$$

Eliminating T_u:

$$\tan\alpha = \sqrt{\frac{f_{yv} A_{stl} u_{cor}}{f_y A_{stl} s}} = \sqrt{\frac{1}{\zeta}} \tag{7-16}$$

Thus

$$T_u = 2A_{cor} \sqrt{\frac{f_y A_{stl} f_{yv} A_{stl}}{u_{cor} s}} = 2\sqrt{\zeta} \frac{f_{yv} A_{stl} A_{cor}}{s} \tag{7-17}$$

$$\zeta = \frac{f_y A_{stl} s}{f_{yv} A_{stl} u_{cor}} \tag{7-18}$$

Where ζ——the reinforcement strength ratio of the longitudinal bars to the stirrups in a torsional member.

When longitudinal reinforcement is asymmetrical, it is calculated according to the symmetrical reinforcement section of the less side. When $\zeta=1$, as is shown in Eq. (7-16), the angle of diagonal strut is about 45°, Eqs. (7-14) and (7-15) can be simplified as:

$$T_u = 2f_y A_{stl} \frac{A_{cor}}{u_{cor}} \tag{7-19}$$

$$T_u = 2f_{yv} A_{stl} \frac{A_{cor}}{s} \tag{7-20}$$

Eqs. (7-19) and (7-20) are based on the space truss model proposed by E. Rausch in 1929.

When ζ is too large or too small, the longitudinal bars or stirrups may not yield. Experimental results show that if α is between 30° and 60° and the corresponding ζ will be between 0.333 and 3 according to Eq. (7-16), the two kinds of steel bars can yield at the member failure provided that the amounts of longitudinal reinforcement and stirrups are appropriate. To limit the crack width of the members under service loads, α should satisfy the following condition:

$$\frac{3}{5} \leqslant \tan\alpha \leqslant \frac{5}{3} \qquad (7\text{-}21)$$

$$\text{or} \quad 0.36 \leqslant \zeta \leqslant 2.778 \qquad (7\text{-}22)$$

As can be seen from the Eq. (7-18), the torsional capacities of the cross-section in pure torsion depends primarily on the size of steel skeleton, the amount of longitudinal reinforcement and stirrups, and relevant yield strength. To avoid compression failure, the maximum amount of reinforcement or the average compressive stress of the diagonal strut σ_c should be limited.

7.3.3 Calculation Method for Torsional Capacities of the Members under Torsion in GB 50010

Code for Design of Concrete Structures GB 50010 integrates the variable angle space truss model and test data with the requirements of reliability. The torsional capacity about rectangular cross-section, box section, T-shaped and I-shaped cross-sections are given (Figure 7-9).

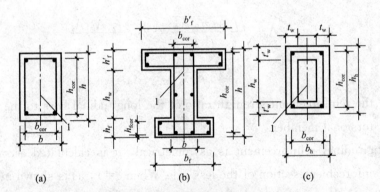

Figure 7-9 Torsion member sections
(a) Rectangular cross section ($h \geqslant b$); (b) T and I-section; (c) Box section ($t_w \leqslant t'_w$)

(1) The equation for torsion resistance of reinforced concrete members of rectangular sections subjected to pure torsion specified in GB 50010 is:

$$T_u = 0.35 f_t W_t + 1.2 \sqrt{\zeta} \frac{f_{yv} A_{st1} A_{cor}}{s} \qquad (7\text{-}23)$$

Where ζ——the reinforcement strength ratio of the longitudinal bars to the stirrups in a torsional member, $0.6 \leqslant \zeta \leqslant 1.7$; if $\zeta > 1.7$, $\zeta = 1.7$ is taken, if $\zeta < 0.6$, $\zeta = 0.6$ is taken;

A_{st1}——the cross-sectional area of one stirrup leg;

f_{yv}——the tensile strength design value of stirrups in twist, $f_{yv} \leqslant 360 \text{N/mm}^2$;

A_{cor}——the area enclosed by the inside profile of the stirrups, $A_{cor} = b_{cor} \cdot h_{cor}$;

u_{cor}——the perimeter of the core part of the section, $u_{cor}=2(b_{cor}+h_{cor})$;

s——the spacing of stirrup in torsion.

Where the first term on the right-hand side of the equation is the torsion resistance of concrete and the second term is the torsion resistance of reinforcement.

For the torsion of reinforcement, it can use the variable angle space truss model to describe. Comparing Eqs. (7-23) and (7-17), the factor in front of the reinforcement space truss analogy is taken as 1.2 in the code while 2.0 in Eq. (7-17). The reasons include: the concrete resistance torque has already been considered by the first term; the calculation of A_{cor} in the code equation is based on the area enclosed by the inside profile of the stirrup, whereas A_{cor} in the space truss model is based on the area enclosed by the connection lines of the centroids of the longitudinal bars at the corners of the section; as shown in Figure 7-10, the factors of 1.2 and 0.35 in Eq (7-23) are on the basis of statistical test data and the reliability index β_c is based on the lower limit of the test points.

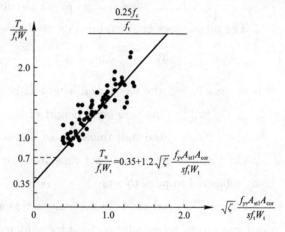

Figure 7-10 Comparison of calculation and experimental data

Experiment shows that when ζ is between 0.5 and 2.0, both the longitudinal bars and the stirrups can yield; meanwhile, the maximum width of cracks can be limited to an allowable value under the service load. GB 50010 conservatively specifies $0.6 \leqslant \zeta \leqslant 1.7$. And if $\zeta > 1.7$, $\zeta = 1.7$ is taken; if $\zeta < 0.6$, $\zeta = 0.6$ is taken.

The equation for torsion resistance of reinforced concrete members of rectangular sections subjected to torsion and axial compression in GB 50010 is:

$$T_u = 0.35 f_t W_t + 1.2\sqrt{\zeta}\frac{f_{yv}A_{st1}A_{cor}}{s} + 0.07\frac{N}{A}W_t \qquad (7-24)$$

Where ζ——calculated by Eq. (7-18), $0.6 \leqslant \zeta \leqslant 1.7$, and if $\zeta > 1.7$, $\zeta = 1.7$ is taken, if $\zeta < 0.6$, $\zeta = 0.6$ is taken;

N——axial compression of design value. When $N > 0.3 f_c A$, $N = 0.3 f_c A$ is taken.

(2) Calculation of torsional capacities for members of box sections subjected to pure torsion

Both experimental results and theoretical study show that when the section width, the height of the concrete strength and the reinforcement are exactly the same, torsion capacities for box-section member with certain wall thickness and solid section member are the same. For the box-section members in pure torsion, the first term on the right-hand side of the equation is multiplied by the reduction factor relative with the wall thickness in GB 50010, namely:

$$T_u = 0.35\alpha_h f_t W_t + 1.2\sqrt{\zeta}\frac{f_{yv}A_{stl}A_{cor}}{s} \quad (7-25)$$

Where α_h —— the influence factor with wall thickness of box-section, $\alpha_h = 2.5t_w/b_h$, b_h is the width of box-section, t_w is the wall thickness of box section with $t_w \geqslant b_h/7$. when $\alpha_h > 1$, $\alpha_h = 1$ is taken.

The plastic torsional modulus of the box-section is:

$$W_t = \frac{b_h^2}{6}(3h_h - b_h) - \frac{(b_h - 2t_w)^2}{6}[3h_w - (b_h - 2t_w)] \quad (7-26)$$

Where b_h, h_h —— the width and height of box-section;

h_w —— the web clear height of box-section;

t_w —— the wall thickness of box-section.

(3) Calculation of torsional capacities for members with T-shaped and I-shaped sections subjected to pure torsion

For the T-shaped and I-shaped cross-section members in torsion, the cross-section can be divided into several rectangular portions for reinforcement calculation. Firstly, the principle for that is to divide the cross section of the web in total height of the cross-section, and then the compression flange and tension flange area can be obtained, as shown in Figure 7-11. The ultimate torque of a cross section composed of several rectangles can be conservatively calculated by summing up the ultimate torques from the component rectangles.

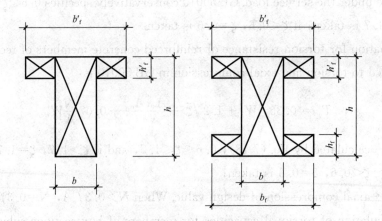

Figure 7-11 Optimal subdivisions of cross sections

1) The torque of the web:

$$T_w = \frac{W_{tw}}{W_t}T \quad (7-27)$$

2) The torque of the compression flange:

$$T'_f = \frac{W'_{tf}}{W_t} T \tag{7-28}$$

3) The torque of the tension flange:

$$T_f = \frac{W_{tf}}{W_t} T \tag{7-29}$$

Where T is the torque design value of the entire cross section. T_w is the torque design value of the cross section of web. T_f and T'_f are the torque design values of the compression flange and tension flange. W_{tw}, W'_{tf}, W_{tf} and W_t are the plastic torsion modulus of the web, compression flange, tension flange and the total cross section, which can be figured out as:

$$W_{tw} = \frac{b^2}{6}(3h-b) \tag{7-30}$$

$$W'_{tf} = \frac{h'^2_f}{2}(b'_f - b) \tag{7-31}$$

$$W_{tf} = \frac{h^2_f}{2}(b_f - b) \tag{7-32}$$

The total cross section of the plastic torsion modulus is:

$$W_t = W_{tw} + W'_{tf} + W_{tf} \tag{7-33}$$

For the above equation, the flange width should meet $b'_f \leqslant b + 6h'_f$ and $b_f \leqslant b + 6h_f$.

7.4 Bearing Capacity of Members Under Combined Torsion, Shear and Flexure

7.4.1 Failure Modes

The stress states in members under the combined action of bending moment M, shear force V and torque T are very complicated. The failure modes and load-carrying capacities of members are related to the load effect and internal factors of members. The torsion-moment ratio $\psi = T/M$ and torsion-shear ratio $\chi = T/(Vb)$ are adopted for the load effect. The internal factors of members include the section size of components, reinforcement and strength of materials. Research has indicated that the members will fail in the following three modes.

1. Bending failure mode (Figure 7-12a)

This occurs when the torque-moment ratio ψ is small. The inclined crack is firstly observed on the top surface, then propagates on the bottom surface and to the side faces. No crack will be observed under flexural compression. At failure, both the longitudinal bars and the stirrups that intersect with the spiral cracks yield in tension. The member's top section is under compression.

2. Torsion failure mode (Figure 7-12b)

This occurs when both the torsion-moment ratio ψ and the torsion-shear ratio χ are large with the longitudinal reinforcement at the top less than that at the bottom. Because the bending moment is small, the flexure-induced compressive stress at the member's upper section is not sufficient to counteract the torsion-induced tensile stress. Longitudinal bars at both the top and the bottom of the member are in tension. Furthermore, because the longitudinal reinforcement is not symmetrically provided, the tensile stresses in the top bars are even larger than those in the bottom bars. Therefore, torsional inclined cracks first appear at the top of the member and spread to the side surfaces, whereas bottom of the section remains in compression.

3. Shear-torsion failure mode (Figure 7-12c)

This occurs when the bending moment is small and the shear and torque are dominant. As this mode of failure occurs, both the shear force and the torque cause shear stress in the cross section. The superposition of the shear stress from different sources leads to the shear stress increasing on one section side and decreasing on the opposite side. Therefore, torsional inclined cracks first appear on the side surface with larger shear stress and then propagate to the top and bottom surfaces. The other side surface is in compression. At failure, both the longitudinal bars and the stirrups that intersect with the spiral cracks yield.

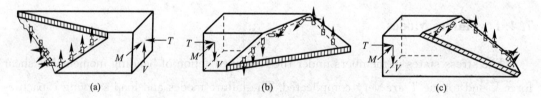

Figure 7-12 The failure modes of members under combined torsion, shear and moment.
(a) Bending failure mode; (b) Torsion failure mode; (c) Shear-torsion failure mode

As mentioned in Chapter 4, without the torque, the shear-compressive failure may happen in the inclined section. For the members under combined torsion, shear and moment, in addition to the above-mentioned three failure modes, shear-type failure may also occur if the shear force is large but the torque is small, which is similar to the shear-compression failure.

The calculation of the members under combined torsion, shear and moment is also based on the variable angle space truss model analogy and the skew bending theory.

7.4.2 Method of Reinforcement Calculation According to GB 50010

For the members under combined torsion, shear and moment, the calculation is very complicated when using the variable angle space truss model and the skew bending theory. On the basis of large number of experimental results in China and the variable angle space truss model, GB 50010 gives the practical reinforcement calculation method about shear-torsion, flexure-torsion and flexure-shear-torsion.

The formulas of shear and torsion capacity in GB 50010 should consider the role of concrete. Hence, the calculation equation of shear-torsion members should integrate the influence of the torque and the shear into the torsion and shear load-carrying capacity. According to the different sections, equations are similar to the load-carrying capacity of pure torsion members in GB 50010.

1. Bearing capacities of members under combined torsion and shear of rectangular cross-sections

The shear load-carrying capacity of shear-torsion members is:

$$V_u = 0.7(1.5 - \beta_t) f_t b h_0 + f_{yv} \frac{A_{sv}}{s} h_0 \tag{7-34}$$

The torsion load-carrying capacity of shear-torsion members is:

$$T_u = 0.35 \beta_t f_t W_t + 1.2 \sqrt{\zeta} f_{yv} \frac{A_{stl} A_{cor}}{s} \tag{7-35}$$

Where β_t——The torsional capacity reduction factor for concrete.

$$\beta_t = \frac{1.5}{1 + 0.5 \frac{V}{T} \frac{W_t}{b h_0}} \tag{7-36}$$

For independent members of rectangular cross sections under concentrated loads, the calculation equation of shear capacity is:

$$V_u = \frac{1.75}{\lambda + 1}(1.5 - \beta_t) f_t b h_0 + f_{yv} \frac{A_{sv}}{s} h_0 \tag{7-37}$$

The torsional capacity is the same as (7-35), but β_t equals:

$$\beta_t = \frac{1.5}{1 + 0.2(\lambda + 1) \frac{V}{T} \frac{W_t}{b h_0}} \tag{7-38}$$

If $\beta_t < 0.5$, the influence of torque moment on the shear capacity can be neglected, and $\beta_t = 0.5$ is taken. If $\beta_t > 1.0$, the influence of shear moment on the torsion capacity can be neglected, and $\beta_t = 1.0$. λ is the shear span ratio of the calculation section.

2. Bearing capacities of members under combined torsion and shear of box sections

The shear load-carrying capacity of shear-torsion members is

$$V_u = 0.7(1.5 - \beta_t) f_t b h_0 + f_{yv} \frac{A_{sv}}{s} h_0 \qquad (7\text{-}39)$$

The torsion load-carrying capacitiy of shear-torsion members is:

$$T_u = 0.35 \alpha_h \beta_t f_t W_t + 1.2 \sqrt{\zeta} f_{yv} \frac{A_{stl} A_{cor}}{s} \qquad (7\text{-}40)$$

Where α_h and ζ should be based on the torsion load-carrying capacities of box section reinforced concrete in pure torsion.

For box section, W_t is replaced by $\alpha_h W_t$ in Eq. (7-36)

$$\beta_t = \frac{1.5}{1 + 0.2(\lambda + 1) \dfrac{V \alpha_h W_t}{T b_h h_0}} \qquad (7\text{-}41)$$

For independent members of box sections under concentrated loads, the equation of the shear load-carrying capacity is same with Eq. (7-37), but β_t should be calculated according to Eq. (7-41).

For independent members of box sections under concentrated loads, the equation of the shear torsion capacitiy is same with Eq. (7-40), but W_t is replaced by $\alpha_h W_t$ and β_t should be calculated according to Eq. (7-41).

3. Bearing capacities of members under combined torsion and shear of T-shaped and I-shaped sections

The shear capacities of members under combined torsion and shear can be calculated according to the Eqs. (7-34) and (7-36) or Eqs. (7-37) and (7-38), but T and W_t are replaced by T_w and W_{tw}, which is supposed that the shear is borne by the web.

The torsion capacities of members under combined torsion and shear can be calculated according to the calculation of the members in pure torsion and the section is divided into several rectangular sections. The web is the shear-torsion member, calculated by Eqs. (7-35) and (7-36) or Eq. (7-38) with T and W_t replaced by T_w and W_{tw}. The tension flange and compression flange is based on pure torsion members according to the rectangular cross sections with T and W_t replaced by T'_f, W'_{tf} and T_f, W_{tf}.

The calculation of rectangular cross sections, T-shape, I-shape and box sections should obey the following principles in general: the reinforced area and the corresponding position with reinforcement of longitudinal reinforcement should be arranged separately according to the flexural capacities of the normal section members in flexure and the torsion capacities of the shear-torsion members. Stirrups should be configured separately according to the shear capacities of the shear-torsion members and the torsion capacities.

Therefore, for rectangular cross-section members under flexure, shear and torsion, when the M, V and T are known, the longitudinal reinforcement (A_s and A'_s) can be calculated according to M, which is based on the capacities of the normal section members in pure flexure. β_t is calculated according to V and T, which is based on Eq. (7-36) or (7-38). Eqs. (7-34) and (7-35) or Eqs. (7-37) and (7-35) are to calculate the stirrups area (A_{sv}) subjected to shear and the longitudinal reinforcement (A_{stl}) and stirrups (A_{st1}) subjected to torsion. Then the reinforcements can be configured in the corresponding position.

In GB 50010, the calculation for capacities of members under combined torsion (T), shear (V) and moment (M), among which T and V are smaller, should be based on the following equation.

(1) When $V \leqslant 0.35 f_t b h_0$ or $V \leqslant 0.875 f_t b h_0 /(\lambda+1)$, the shear can be neglected. Only moment capacities of the normal section in pure flexure and the torsion capacities of the twist section in pure torsion are considered.

(2) When $T \leqslant 0.175 f_t W_t$ or $T \leqslant 0.175 \alpha_h f_t W_t$ of box section, the torque can be neglected. Only the moment capacities of the normal section in pure flexure and the shear capacities of the inclined section are considered.

4. The torsional capacity reduction factor β_t for concrete

The relationship between shear and torsion capacities is adopted to obtain β_t from Eq. (7-36) or Eq. (7-38).

Test results of concrete members under combined torsion and shear show that there is a correlative relationship in the form of 1/4 circle between the shear strength and the torsion strength as shown in Figure 7-13(a). V_{co} is the shear resistance of a concrete member under pure shear and T_{co} is the torsion resistance of a concrete member under pure torsion. In GB 50010, the 1/4 circle relationship is substituted by the three-straight line segments as shown in Figure 7-13(b).

When

$$\frac{V_c}{V_{co}} \leqslant 0.5, \quad \frac{T_c}{T_{co}} = 1.0 \tag{7-42}$$

When

$$\frac{T_c}{T_{co}} \leqslant 0.5, \quad \frac{V_c}{V_{co}} = 1.0 \tag{7-43}$$

When

$$\frac{T_c}{T_{co}}, \frac{V_c}{V_{co}} > 0.5, \quad \frac{T_c}{T_{co}} + \frac{V_c}{V_{co}} = 1.5 \tag{7-44}$$

For the Eq. (7-44)

$$\text{If } \frac{T_c}{T_{co}} = \beta_t \tag{7-45}$$

Then $\frac{V_c}{V_{co}} = 1.5 - \beta_t$

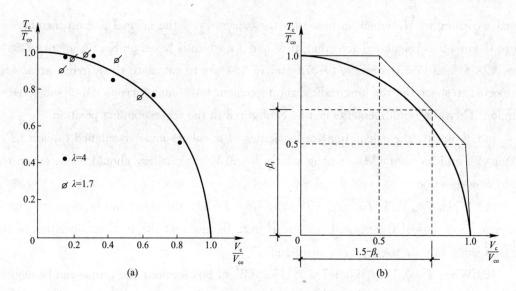

Figure 7-13 Correlative relation under combined torsion and shear

(a) Correlative relation of a quarter circle; (b) Correlative relation of three-folded line

$$\text{Thus,} \quad \beta_t = \frac{1.5}{1 + \frac{V_c/V_{co}}{T_c/T_{co}}} \tag{7-46}$$

Where V_c and T_c are the shear and torsion load-carrying capacities of web reinforcement members under combined torsion and shear. T_{co} and V_{co} are the torsion and shear capacities of web reinforcement members under pure torsion and pure shear, respectively. In the Eq. (7-46), $\frac{V_c}{T_c}$ is replaced by $\frac{V}{T}$, $T_{co} = 0.35 f_t W_t$ and $V_{co} = 0.7 f_t bh_0$ are taken, then Eq. (7-36) can be obtained.

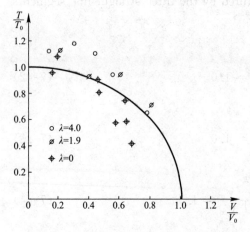

Figure 7-14 The relation of shear load-carrying capacity and torsion load-carrying capacity for web reinforcement members

$T_{co} = 0.35 f_t W_t$ and $V_{co} = \frac{1.75}{\lambda + 1} f_t bh_0$ are taken to obtain Eq. (7-38).

Test results of the web reinforcement members under combined torsion, shear and flexure show that there is a correlative relationship in the form of a quarter circle, as is shown in Figure 7-14, T_0 is the torsion capacity of web reinforcement members under pure torsion, and V_0 is the shear capacity of web reinforcement members when the torque equals to zero with different λ values. According to the calculation of aforementioned variable angle space truss model, although the correlative relationship about the shear-torsion capacities is very complex, the relationship between shear capacity and torsion capacity of members generally can be described by 1/4 circle curve approximately.

7.5 Calculation for Torsional Capacities of Reinforced Concrete Frame Column Members with Rectangular Section under Combined Torsion, Shear, Flexure and Axial Force

1. Axial force in compression

According to GB 50010, the equation of shear-torsional capacities of reinforced concrete frame column members with rectangular section under combined torsion, shear, flexure and axial compression is:

(1) Capacities of shear

$$V_u = (1.5 - \beta_t)\left(\frac{1.75}{\lambda+1}f_t bh_0 + 0.07N\right) + f_{yv}\frac{A_{sv}}{s}h_0 \qquad (7\text{-}47)$$

(2) Capacities of torsion

$$T_u = \beta_t\left(0.35f_t W_t + 0.07\frac{N}{A}W_t\right) + 1.2\sqrt{\zeta}f_{yv}\frac{A_{stl}A_{cor}}{s} \qquad (7\text{-}48)$$

Where β_t——The calculation variable based on Eq. (7-38).

For the reinforced concrete frame columns with rectangular section under combined torsion, shear, flexure and axial compression, the reinforced area of longitudinal reinforcement should be calculated according to the capacities of the normal section in eccentric compression members and the torsion capacities of shear-torsion members. Stirrups should be configured separately according to the shear capacities and the torsion capacities of the shear-torsion members.

In cross-sectional design, when $T \leqslant \left(0.175f_t + 0.035\frac{N}{A}\right)W_t$, the influence of the torque can be neglected.

2. Axial force in tension

The equation of shear-torsional capacities of reinforced concrete frame columns with rectangular section under combined torsion, shear, flexure and axial tension is:

(1) Capacities of shear

$$V_u = (1.5 - \beta_t)\left(\frac{1.75}{\lambda+1}f_t bh_0 - 0.2N\right) + f_{yv}\frac{A_{sv}}{s}h_0$$

When

$$V_u < f_{yv}\frac{A_{sv}}{s}h_0, \ V_u = f_{yv}\frac{A_{sv}}{s}h_0 \text{ is taken} \qquad (7\text{-}49)$$

(2) Capacities of torsion

$$T_u = \beta_t \left(0.35 f_t - 0.2 \frac{N}{A}\right) W_t + 1.2 \sqrt{\zeta} f_{yv} \frac{A_{stl} A_{cor}}{s} \tag{7-50}$$

When

$$T_u < 1.2 \sqrt{\zeta} f_{yv} \frac{A_{stl} A_{cor}}{s}, \quad T = 1.2 \sqrt{\zeta} f_{yv} \frac{A_{stl} A_{cor}}{s} \text{ is taken}$$

Where N——the axial compression or axial tension corresponding to shear(V) and torque (T) design value;

λ——the shear-span ratio of the calculated cross section;

A_{sv}——the sectional area of stirrups for shear capacity;

A_{stl}——the cross-section area of one stirrup leg along the cross section in torsion.

When $T \leqslant \left(0.175 f_t - 0.1 \frac{N}{A}\right) W_t$, only the eccentric tension component load-carrying capacity of normal section and inclined section load-carrying capacity are to be calculated.

7.6 Bearing Capacity of Reinforced Concrete Members under Coordination Twist

After the reinforced concrete members cracking, the torsion rigidity drops because of the distribution of internal force. In general circumstances, to simplify the calculation, the torsion rigidity is taken as zero, ignoring the role of torque. However, measures should be configured according to the structure of the twisted longitudinal reinforcement and stirrups to ensure sufficient ductility and to restrain the crack width under service. This is the design method of zero rigidity in some foreign codes. This method isn't taken in GB 50010, but the distribution of internal force is considered, reducing the deign value of torque T, and calculating the capacity according to the combined members.

7.7 Detailing Requirements of the Members under Torsion

1. Detailing requirements of torsional longitudinal reinforcement

(1) To avoid the tension failure, the ratio ρ_{tl} of torsional longitudinal reinforcement should not be less than the minimum ratio $\rho_{stl, min}$.

$$\rho_{tl} = \frac{A_{stl}}{bh} \geqslant \rho_{stl, min}$$

$$\rho_{stl, min} = \frac{A_{stl, min}}{bh} = 0.06 \sqrt{\frac{T}{Vb}} \frac{f_t}{f_y} \tag{7-51}$$

When $\frac{T}{Vb} > 2$, $\frac{T}{Vb} = 2$ is taken.

(2) The space of torsional longitudinal reinforcement should not be more than 200mm and the width of the beam.

(3) At the four corners of the cross-section, torsional longitudinal reinforcement should be arranged, which are symmetrically along the perimeter of cross section; when there is a greater torque effect at the support, the longitudinal torsional bars should be anchored at the support according to the sufficient tension.

(4) In the members under combined flexure, shear and torsion, the longitudinal bars are configured by the bending section in tension. The areas should not be less than the sum of the section areas calculated by the minimum ratio of flexural members and torsional longitudinal reinforcement and should be assigned to the side of flexure-tension.

2. Detailing requirements of torsional stirrups

(1) To avoid the tension failure, the ratio ρ_{sv} of torsional stirrups should not be less than the minimum ratio of $0.28 \frac{f_t}{f_{yv}}$ in the members under combined flexure, shear and torsion:

$$\rho_{sv} = \frac{nA_{sv1}}{bs} \geqslant 0.28 \frac{f_t}{f_{yv}} \tag{7-52}$$

(2) The principal tensile stresses caused by the torque exist on all sides of the members, so the torsion stirrups should be closed up into a hoop with both ends connected preferably by welding.

(3) The torsional stirrups should have both ends bent through a 135° hook embedded in the core of the members with a length not less than 10 times the diameter of the stirrup.

Longitudinal torsional bars are acting as upper and lower chords, concrete is diagonal web chord and stirrups are vertical webs in the variable angle space truss model. Hence, both stirrups in torsion and longitudinal reinforcement in torsion should function cooperatively at the same time.

For the box-section members, b is replaced by b_h in Eqs. (7-51) and (7-52) and the meaning of b_h is show in Figure 7-9.

3. Detailing requirements for dimensions of members

In order to avoid the over-reinforced damage of members under flexure-shear-torsion, GB 50010 specifies that the h_w/b should not be more than 6 for rectangular section, T-shaped section and I-shaped section, and h_w/t_w should not be more than 6 for box sections. The dimensions of members should be conformed in the following conditions:

When h_w/b is not more than 4

$$\frac{V}{bh_0} + \frac{T}{0.8w_t} \leqslant 0.25\beta_c f_c \tag{7-53}$$

When h_w/b equals to 6

$$\frac{V}{bh_0} + \frac{T}{0.8w_t} \leqslant 0.2\beta_c f_c \tag{7-54}$$

When $4 < h_w/b\,(h_w/t_w) < 6$, the value is determined by liner interpolation.

When $6 < h_w/b\,(h_w/t_w)$, the requirements of the dimensions and the calculation for load-carrying capacity of twist section should confirm the specific requirements for the torsional members.

Where T——the design value of torque;

$\quad\quad b$——the width of the rectangular section; for T-shape and I-shape sections, b is the width of the web; for the box section, b is the total thickness of the side walls, namely $2t_w$;

$\quad\quad w_t$——the sectional plastic torsion modulus for torsion members;

$\quad\quad h_w$——the height of web in section; for rectangle section, the effective height h_0 is taken; for T-shape section, the effective height minus the height of the flange is taken; for I-shape section, the clear height is taken;

$\quad\quad t_w$——the wall thickness of box-section, the value should not be less than $b_h/7$, where b_h is the width of the box-section.

4. Configuration about stirrups in torsion and torsional longitudinal reinforcement in detailing requirements

For the members under combined moment, shear and torque, the stirrups in torsion and torsional longitudinal reinforcement according to the structure requirements are required in the following conditions.

$$\frac{V}{bh_0} + \frac{T}{W_t} \leqslant 0.7 f_t \tag{7-55}$$

Or

$$\frac{V}{bh_0} + \frac{T}{W_t} \leqslant 0.7 f_t + 0.07 \frac{N}{bh_0} \tag{7-56}$$

Where N——the design value of axial compression corresponding to shear (V) and torque (T) design values. When $N > 0.3 f_c A$, $N = 0.3 f_c A$ is taken.

[**Example 7-1**]   A T-shape section of a beam under the combined actions of a design torque $T = 15 \text{kN} \cdot \text{m}$, bending moment $M = 110 \text{kN} \cdot \text{m}$ and shear force $V = 120 \text{kN}$ by uniformly distributed load is designed; the dimensions of section are $b \times h = 250\text{mm} \times 500\text{mm}$, $b_f' = 400\text{mm}$, $h_f' = 100\text{mm}$; the concrete is of grade C30; the longitudinal reinforcement is HRB400 and HRB335 is designed for the stirrup. The member is under Class 1 environmental condition.

The reinforcements respective for flexure, shear and torsion are to be designed.

[Solution]

$f_c = 14.3\text{N/mm}^2$, $f_t = 1.43\text{N/mm}^2$, $f_y = 360\text{N/mm}^2$, $f_{yv} = 300\text{N/mm}^2$

(1) the section dimensions of members are to be examined

$$h_0 = h - a_s = 500 - 40 = 460\text{mm}$$

$$W_{tw} = \frac{b^2}{6}(3h - b) = \frac{250^2}{6} \times (3 \times 500 - 250) = 1302.1 \times 10^4 \text{mm}^3$$

$$W'_{tf} = \frac{h'^2_f}{2}(b'_f - b) = \frac{100^2}{2} \times (400 - 250) = 75 \times 10^4 \text{mm}^3$$

$$W_t = W_{tw} + W'_{tf} = (1302.1 + 75) \times 10^4 \text{mm}^3 = 1377.1 \times 10^4 \text{mm}^3$$

According to $\dfrac{V}{bh_0} + \dfrac{T}{0.8W_t} \leqslant 0.25\beta_c f_c$ and $\dfrac{V}{bh_0} + \dfrac{T}{W_t} \leqslant 0.7f_t$

$$\frac{V}{bh_0} + \frac{T}{0.8W_t} = \frac{120 \times 10^3}{250 \times 465} + \frac{15 \times 10^6}{0.8 \times 1377.1 \times 10^4}$$

$$= 2.39\text{N/mm}^2 \leqslant 0.25\beta_c f_c = 0.25 \times 1.0 \times 14.3 = 3.58\text{N/mm}^2$$

$$\frac{V}{bh_0} + \frac{T}{W_t} = \frac{120 \times 10^3}{250 \times 465} + \frac{15 \times 10^6}{1377.1 \times 10^4}$$

$$= 2.12\text{N/mm}^2 > 0.7f_t = 0.7 \times 1.43 = 1.0\text{N/mm}^2$$

The section dimensions meet the requirements, the followings are to design the reinforcements.

(2) Selection for the method of calculation

$$T = 15\text{kN} \cdot \text{m} > 0.175 \times f_t W_t = 0.175 \times 1.43 \times 1377.1 \times 10^4 = 3.45\text{kN} \cdot \text{m}$$

$$V = 120\text{kN} > 0.35 f_t bh_0 = 0.35 \times 1.43 \times 250 \times 456 = 58.18\text{kN}$$

Thus, the torsion and shear impacts on the shear and torsion capacities should be considered.

(3) Calculation for the reinforcement in flexure

According to

$$\alpha_1 f_c b'_f h'_f \left(h_0 - \frac{h'_f}{2}\right)$$

$$= 1.0 \times 14.3 \times 400 \times 100 \times \left(460 - \frac{100}{2}\right) = 234.52\text{kN} \cdot \text{m} > 110\text{kN} \cdot \text{m}$$

The member belongs to the first category of T-beams

$$\alpha_s = \frac{M}{\alpha_1 f_c b'_f h_0^2} = \frac{110 \times 10^6}{1.0 \times 14.3 \times 400 \times 460^2} = 0.091$$

Thus, $\gamma_0 = 0.5(1 + \sqrt{1 - 2\alpha_s}) = 0.5 \times (1 + \sqrt{1 - 2 \times 0.091}) = 0.952$

$$A_s = \frac{M}{f_y \gamma_0 h_0} = \frac{120 \times 10^6}{360 \times 0.952 \times 460} = 761 \text{mm}^2$$

(4) Calculation for the reinforcement in shear and torsion

1) The torque of the web and flange in compression

The web:
$$T_w = \frac{W_{tw}}{W_t} T = \frac{1302.1 \times 10^4}{1377.0 \times 10^4} \times 15 \times 10^6 = 14.18 \text{kN} \cdot \text{m}$$

The flange in compression:
$$T'_f = \frac{W'_{tf}}{W_t} T = \frac{75 \times 10^4}{1377.1 \times 10^4} \times 15 \times 10^6 = 0.817 \text{kN} \cdot \text{m}$$

2) Calculation for the reinforcement of the web
$$A_{cor} = b_{cor} \times h_{cor} = 194 \times 444 = 86136 \text{mm}^2$$
$$u_{cor} = 2(b_{cor} + h_{cor}) = 2 \times (194 + 444) = 1276 \text{mm}$$

① Calculation for the stirrups in torsion

According to Eq. (7-36)
$$\beta_t = \frac{1.5}{1 + 0.5 \dfrac{V}{T_w} \dfrac{W_{tw}}{bh_0}} = \frac{1.5}{1 + 0.5 \dfrac{120 \times 10^3}{14.18 \times 10^6} \times \dfrac{1302.1 \times 10^4}{250 \times 460}} = 1.014 > 1.0$$

$$\beta_t = 1.0$$

$\zeta = 1.2$ is taken, $T = T_u$, according to Eq. (7-35)

$$\frac{A_{st1}}{s} = \frac{T_w - 0.35 \beta_t f_t W_{tw}}{1.2 \sqrt{\zeta} f_{yv} A_{cor}}$$

$$= \frac{14.18 \times 10^6 - 0.35 \times 1.0 \times 1.43 \times 1302.1 \times 10^4}{1.2 \sqrt{1.2} \times 300 \times 9.0 \times 10^4}$$

$$= 0.216 \text{mm}^2/\text{mm}$$

The calculation of stirrups in shear is obtained according to Eq. (7-34):

$$\frac{A_{sv}}{s} = \frac{V_u - 0.7(1.5 - \beta_t) f_t bh_0}{1.25 f_{yv} h_0}$$

$$= \frac{120 \times 10^3 - 0.7 \times (1.5 - 1.0) \times 1.43 \times 250 \times 460}{1.25 \times 300 \times 460}$$

$$= 0.362 \text{mm}^2/\text{mm}$$

The total area of single-limb stirrup in the web:

$$\frac{A_{st1}}{s} + \frac{A_{sv}}{2s} = 0.216 + \frac{0.362}{2} = 0.397 \text{mm}^2/\text{mm}$$

The diameter of stirrups is taken as 8mm, so the area of one single leg is 50.3mm², and the space is:

$$s = \frac{50.3}{0.397} = 126.7 \text{mm}, \quad s = 120 \text{mm is taken}$$

② Calculation for the reinforcement bars in torsion

According to Eq. (7-18)

$$A_{stl} = \frac{\zeta f_{yv} A_{stl} u_{cor}}{f_y \cdot s} = \frac{1.2 \times 300 \times 0.216 \times 1276}{360} = 275.6 \text{mm}^2$$

The area of longitudinal reinforcement under flexural and torsion at the bottom surface of the web section is:

$$A_s + A_{stl} \frac{(b_{cor} + 2 \times 0.25 h_{cor})}{u_{cor}}$$

$$= 761 + 275.6 \times \frac{194 + 0.5 \times 444}{1276} = 805.85 \text{mm}^2$$

The diameter is chosen as 20mm and the total area is 942mm² with three bars.

The area of torsional longitudinal reinforcement on both sides of the web in the cross-section is:

$$A_{stl} \frac{2 \times 0.25 h_{cor}}{u_{cor}} = 275.6 \times \frac{444}{1276} = 98.9 \text{mm}^2$$

The diameter is chosen as 12mm, and the total area is 226.1mm² with two bars.

The area of torsional longitudinal reinforcement on top of the web in the cross-section is:

$$A_{stl} \frac{(b_{cor} + 2 \times 0.25 h_{cor})}{u_{cor}} = 275.6 \times \frac{194 + 0.5 \times 444}{1276} = 89.85 \text{mm}^2$$

The diameter is chosen as 12mm with the total area is 226mm² of two bars.

3) Calculation for the reinforcement of the compression flange

$$A'_{cor} = b'_{cor} \times h'_{cor} = 94 \times 44 = 4136 \text{mm}^2$$

$$u'_{cor} = 2(b'_{cor} + h'_{cor}) = 2 \times (94 + 44) = 276 \text{mm}$$

① Calculation for the stirrups in torsion

$\zeta = 1.0$, according to Eq. (7-23)

$$\frac{A'_{stl}}{s} = \frac{T'_f - 0.35 f_t W'_{tf}}{1.2 \sqrt{\zeta} f_{yv} A'_{cor}} = \frac{0.817 \times 10^6 - 0.35 \times 1.43 \times 75 \times 10^4}{1.2 \times \sqrt{1.0} \times 300 \times 4136}$$

$$= 0.297 \text{mm}^2/\text{mm}$$

The diameter is chosen as 8mm with two-legged stirrups, the area for one single leg is:

$$A'_{stl} = 50.3 \text{mm}^2$$

Thus, the space of stirrups is:

$$s = \frac{50.3}{0.297} = 169 \text{mm}, \quad s = 150 \text{mm is taken}$$

② Calculation for the reinforcement in torsion

According to Eq. (7-18)

$$A_{stl} = \frac{\zeta f_{yv} A'_{stl} u'_{cor}}{f_y} = \frac{1.0 \times 300 \times 0.297 \times 276}{360} = 68.31 \text{mm}^2$$

The diameter is chosen as 8mm with 4 bars and the total area is 201mm².

(5) the minimum ratio of stirrups in the web

According to Eq. (7-52)

$$\rho_{sv,min} = 0.28 \frac{f_t}{f_{yv}} = 0.28 \times \frac{1.43}{300} = 0.0013$$

In fact

$$\rho_{sv} = \frac{nA_{svl}}{bs} = \frac{2 \times 50.3}{250 \times 120} = 0.0034 > 0.0013$$

(6) the ratio of reinforcement in the web under torsion

According to Eq. (7-51)

$$\rho_{stl,min} = \frac{A_{stl,min}}{bh} = 0.6\sqrt{\frac{T_w}{Vb}}\frac{f_t}{f_y}$$

$$= 0.6 \times \sqrt{\frac{14.18 \times 10^6}{120 \times 10^3 \times 250}} \times \frac{1.43}{360} = 0.0016$$

The minimum ratio of reinforcement in torsion members

$$\rho_{s,min} = 0.45 \frac{f_t}{f_y} = 0.45 \times \frac{1.43}{360} = 0.178\% < 0.2\%$$

$\rho_{s,min} = 0.2\%$ is taken.

The minimum ratio of longitudinal reinforcement in the section of the tension side for bending

$$\rho_{s,min}bh_0 + \rho_{stl,min}bh \frac{(b_{cor} + 2 \times 0.25 h_{cor})}{U_{cor}}$$

$$= 0.002 \times 250 \times 444 + 0.0016 \times 250 \times 500 \times \frac{(194 + 0.5 \times 444)}{1276}$$

$$= 262.75 \text{mm}^2 < 942 \text{mm}^2$$

(7) The ratios for stirrups and reinforcement are sufficient. The longitudinal reinforcements and the stirrups are shown in Figure 7-15.

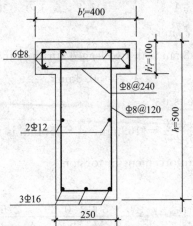

Figure 7-15 The longitudinal reinforcements and the stirrups

Questions

7-1 What are the basic assumptions of variable angle space truss analogy?

7-2 What is ζ? Why is the value of ζ limited? What is the influence of ζ on failure modes?

7-3 What are the failure modes and characteristics for members under pure torsion?

7-4 What is the physical meaning of β_t in the equation for calculating torsional capacity? What is the meaning of β_t in the expression? What factors are considered in this equation?

7-5 What should be taken into consideration in the placement of longitudinal bars and stirrups in torsional members?

7-6 What are the features of the flexure-torsion interaction curve? How will a curve vary with the reinforcement of longitudinal bars?

Exercises

7-1 For a rectangular cross-sectional member under pure torsion, the cross-sectional dimensions are $b \times h = 300 \times 500$mm. The longitudinal bars are $4 \, \Phi \, 14 (f_y = 300\text{N/mm}^2)$ and stirrups are $8@150 (f_y = 270\text{N/mm}^2)$. The concrete is C25 ($f_c = 11.9\text{N/mm}^2$, $f_t = 1.27\text{N/mm}^2$). The maximum torque for the cross section is required.

7-2 For a reinforced concrete member under combined torsion and flexure, the cross-sectional dimensions are $b \times h = 200\text{mm} \times 400\text{mm}$. The bending moment is $M = 70\text{kN} \cdot \text{m}$ and the torque is $T = 12\text{kN} \cdot \text{m}$. C30 concrete is used. The stirrups are HRB335 and the longitudinal bars are HRB400. The reinforcement is required.

7-3 For a member of T-section subjected to combined torsion, shear and flexure (shear force is mainly caused by uniformly distributed loads), the cross-sectional dimensions are $b'_f = 400$mm, $h'_f = 80$mm, $b = 200$mm and $h = 450$mm. The reinforcement of this section is shown in Figure 7-16. The internal forces include a bending moment of $M = 54\text{kN} \cdot \text{m}$, a shear force of $V = 42\text{kN} \cdot \text{m}$ and a torque of $T = 8\text{kN} \cdot \text{m}$. The concrete is C20, and the reinforcement is of HPB300. Whether the cross section can resist the given internal forces is to be figured out ($a_s = 35$mm).

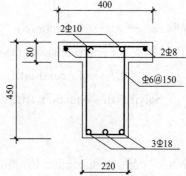

Figure 7-16 Exercise 7-3

Chapter 8
Deflection, Crack and Durability

8.1 Deflection of Reinforced Concrete Members

8.1.1 Rigidity of Flexural Members

Deflection of reinforced concrete flexural members can be obtained by relevant equations in the mechanics of materials class when the flexural stiffness has been determined. The calculation of flexural stiffness should be reasonable and can reflect the plasticity of members after cracking.

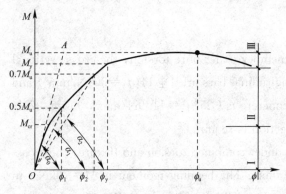

Figure 8-1 M-f curve of a flexural beam

The flexural stiffness of a homogeneous elastic beam is a constant value if the materials and dimensions have been selected. Deflection is linearly proportional to bending moment M, as shown by the dashed line OA in Figure 8-1.

The differential equation for the deflection curve of a homogeneous elastic beam is:

$$\frac{d^2 y(x)}{dx^2} = \frac{1}{r} = -\frac{M(x)}{EI} \qquad (8\text{-}1)$$

Where $y(x)$——deflections;

r——radius of curvature;

EI——flexural stiffness.

Solving this equation, the deflection can be obtained as:

$$f = S \frac{M l_0^2}{EI} \qquad (8\text{-}2)$$

Where M——maximum bending moment;

S——coefficient related to loading types and supporting conditions, for example, $S=5/48$ for simply supported beams under a uniformly distributed load;

l_0 —— computation span.

Under-reinforced concrete beams experience three stages from being loaded to failure (the solid line in Figure 8-1). Before cracking ($M \leqslant M_{cr}$), the beam works elastically. The relationship between f and M is a straight line. If the flexural stiffness is EI, this straight line overlaps the dashed line OA. Once the concrete in the tension zone cracks ($M \geqslant M_{cr}$), the flexural stiffness decreases noticeably. And when the tension reinforcement yields ($M \geqslant M_y$), the flexural stiffness drops sharply. The above phenomena indicate that the flexural stiffness of a reinforced concrete beam is not constant and will be significantly affected by cracking. Because flexural members in service are working with cracks, stage II should be the basis for deformation or deflection calculation.

8.1.2 Short-Term Flexural Stiffness B_s

The bending stiffness of the section not only decreases with the increase of bending moments (or loads), but also decreases with the increase of loading time. The short-term flexural stiffness, which does not consider the loading time, is recorded as B_s.

1. The basic expression of B_s

The Figure 8-2 shows the strain of reinforcement and concrete in the pure bending segment ($M_k = 0.5 M_u^0 \sim 0.7 M_u^0$): (1) Along the length of the beam, the tensile strain of the tensile reinforcements on the cross-section and the compressive strain of the concrete at the edge of the compression section are all unevenly distributed, which reach the maximum ε_{sk} and ε_{ck} at the crack section, respectively, and then decrease gradually. The subscript k represents values produced by the standard combination value M_k of the bending moment;

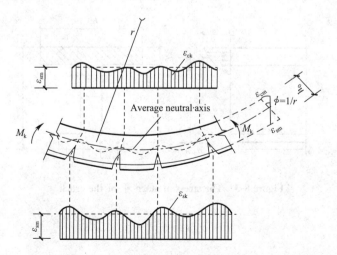

Figure 8-2 Average strain in the pure bending segment

(2) Along the length of the beam, the change of the neutral axis is wavy; (3) When the range of measurement becomes longer ($\geqslant$750mm), the change of the average strain conform to the plane section assumption.

According to the plane section assumption, the curvature of the cross section in a pure bending segment is

$$\varphi = \frac{1}{r} = \frac{\varepsilon_{sm} + \varepsilon_{cm}}{h_0} \tag{8-3}$$

Where r —— the average radius of curvature;

ε_{sm} —— the average tensile strain at the center of gravity in a longitudinal tensile steel bar;

ε_{cm} —— the average compressive strain of the concrete at the edge of the compression zone.

So

$$B_s = \frac{M_k}{\varphi} = \frac{M_k h_0}{\varepsilon_{sm} + \varepsilon_{cm}} \tag{8-4}$$

Where M_k —— standard combination value of the bending moment.

2. The average strain ε_{sm}, ε_{cm}

The average strain ε_{sm} can be expressed by the strain ε_{sk} of the tensile longitudinal reinforcement at the crack section.

$$\varepsilon_{sm} = \psi \varepsilon_{sk} \tag{8-5}$$

Where ψ —— strain inhomogeneous coefficient of longitudinal tensile steel bar between cracks.

Figure 8-3 shows the stress at stage Ⅱ of the crack. The concrete stress in the compression zone can be obtained from the equilibrium conditions as follows:

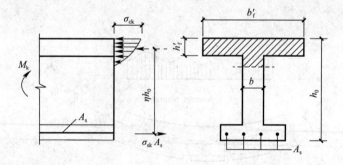

Figure 8-3　The stress at stage Ⅱ of the crack

$$\sigma_{sk} = \frac{M_k}{A_s \eta h_0} \tag{8-6}$$

Where η —— the coefficient of the internal force arm and $\eta = 0.87$.

So

$$\varepsilon_{sm} = \psi \varepsilon_{sk} = \psi \frac{\sigma_{sk}}{E_s} = \psi \frac{M_k}{A_s \eta h_0 E_s} = 1.15 \psi \frac{M_k}{A_s \eta E_s} \qquad (8-7)$$

In addition, according to the experimental study:

$$\varepsilon_{cm} = \frac{M_k}{\xi b E_c h_0^2} \qquad (8-8)$$

In the above formula, ξ is called the composite coefficient of the average strain of concrete at the edge of the compression zone.

3. Strain inhomogeneous coefficient ψ of longitudinal tensile steel bar between cracks

Figure 8-4 shows the strain distribution of the tensile longitudinal reinforcement along the length of the beam. The coefficient ψ reflects the inhomogeneity of tensile steel strain, which represents the performance of tensile concrete in the cracks and reduces the contribution of the concrete to deformation and fracture. The smaller ψ is, the greater the extent of the tensile strength for concrete in the fracture and the greater the contribution to increasing the bending rigidity of the section will be, by which the deformation and the width of the crack will be reduced.

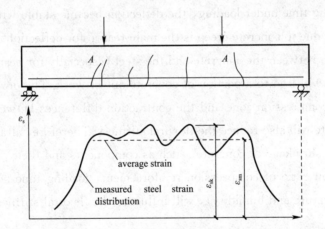

Figure 8-4 Strain distribution of tensile reinforcement in the pure bending segment

The experimental study shows that ψ can be approximately expressed as

$$\psi = 1.1 - 0.65 \frac{f_{tk}}{\rho_{te} \sigma_{sq}} \qquad (8-9)$$

4. The calculation formula of B_s

The experimental data in China and abroad shows that the ξ is related to $\alpha_E \rho$ and the reinforcement coefficient γ_f' of the compression flange. To simplify the calculation, the value of $\alpha_E \rho / \xi$ can be given directly:

$$\frac{\alpha_E \rho}{\xi} = 0.2 + \frac{6 \alpha_E \rho}{1 + 3.5 \gamma_f'} \qquad (8-10)$$

Where

$$\alpha_E = E_s/E_c$$
$$\gamma'_f = (b'_f - b)h'_f/(bh_0)$$

The calculation formula of B_s are obtained by Eqs. (8-5), (8-8), (8-9) and (8-10).

$$B_s = \frac{E_s A_s h_0^2}{1.15\psi + 0.2 + \dfrac{6\alpha_E \rho}{1+3.5\gamma'_f}} \tag{8-11}$$

Where ρ ——The ratio of tensile longitudinal reinforcement.

When $h'_f > 0.2h_0$, h'_f is taken as $h'_f = 0.2h_0$.

8.1.3 Flexural Stiffness B

1. The reasons for the decrease of stiffness under long-term load

Under long-term loads, the flexural stiffness of reinforced concrete members will decrease with time so that the deflection will increase. Experiments have shown that deflections of beams increased quickly in the earlier stage. Then, the increasing rate gradually decreased. After a long time under loading, the deflection became stable with little increase. The increase of ε_{cm} due to concrete creep is the main reason for deflection augment. In addition, the slip creep between the concrete and the steel (especially for beams with small reinforcement ratio) will lead to a portion of concrete out of work. The development of concrete plasticity in compression zone and the contraction differences between tensioned and compressed concrete will also reduce the flexural stiffness. Therefore, all factors that influence the creep and shrinkage of concrete, such as components and their proportion in concrete, reinforcement ratio of compression reinforcement, loading time and environmental conditions (temperature and humidity), will influence the flexural stiffness.

2. Flexural stiffness B

The influence of long-term loads on the increase of deflection can be represented by θ, which is defined as follows:

$$\theta = 2.0 - 0.4 \frac{\rho'}{\rho} \tag{8-12}$$

Where ρ and ρ' ——ratio of tensile and compression reinforcement respectively;

θ ——determined by experiments, considering the restraint of compression reinforcement on creep and shrinkage of concrete.

The deformation is calculated by the characteristic combination of load effects:

$$f = \theta S \frac{M_q l_0^2}{B_s} + S\frac{M_k - M_q}{B_s}l_0^2 = S\frac{M_k}{B}l_0^2 \tag{8-13}$$

Thus, the flexural stiffness calculated by the characteristic combination of load effects and considering the influence of long-term load is

$$B = \frac{M_k}{M_q(\theta-1)+M_k} B_s \quad (8\text{-}14)$$

GB 50010 adopts this calculation method and stipulates that the maximum deflection of reinforced concrete flexural members should be calculated under the quasi-permanent combination of load effects, whereas the maximum deflection of prestressed concrete flexural members is under the characteristic combination of load effects, in both of which the influence of long-term loads should be considered.

8.1.4 Deformation Checking

Currently, research on deformation control is limited to the deflection control of flexural members, as well as trusses and arches in highway bridges. It is generally required that

$$f \leqslant f_{\lim} \quad (8\text{-}15)$$

Deflection limits are mainly determined by the above-mentioned purposes and practical experience.

$$f = S\frac{M_k l_0^2}{B} \quad (8\text{-}16)$$

After the flexural stiffness has been obtained, the deflection of flexural members can be calculated by replacing the flexural stiffness for EI in Eq. (8-12). Then, Eq. (8-13) is used to check whether the deflection is satisfied.

[**Example 8-1**] The hollow plate shown in Figure 8-5(a) has a computation span of $l_0 = 3.04$m. C30 concrete and $9\phi6$ are provided. The concrete cover depth is $c=15$mm. The bending moment calculated under standard combination of load effects is $M_k = 4.47$ kN·m and quasi-permanent combination of load effects is $M_q = 2.91$kN·m. The deformation of the plate is to be checked. $f_{\lim} = l_0/200$.

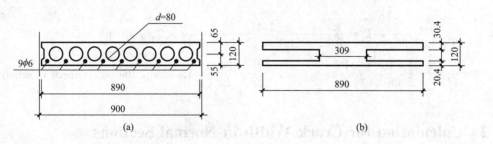

Figure 8-5 Example 8-1 (unit: mm)

[**Solution**]

(1) Calculation for cross section

The cross section of the hollow plate is transformed to an I-shaped section, i.e., the circular holes are replaced by equivalent rectangular holes with the same area, centroid and moment of inertia of the cross section.

$$\frac{\pi d^2}{4} = b_a h_a, \quad \frac{\pi d^4}{64} = \frac{b_a h_a^3}{12}$$

Solving the equation set, $b_a = 72.6$mm and $h_a = 69.2$mm are obtained. The dimensions of the I-shaped section are (Figure 8-5b):

$$b = 890 - 72.6 \times 8 \approx 310\text{mm}, \quad h_0 = 120 - (15+3) = 102\text{mm},$$

$$h_f' = 65 - \frac{69.2}{2} = 30.4\text{mm} > 0.2h_0 = 20.4\text{mm}$$

$$h_f = 55 - \frac{69.2}{2} = 20.4\text{mm}$$

(2) Deflection checking is expressed as follows:

$$\alpha_E \rho = \frac{E_s A_s}{E_c b h_0} = \frac{2.1 \times 10^5}{3.0 \times 10^4} \times \frac{28.3 \times 9}{310 \times 102} = 0.056$$

$$\gamma_f' = \frac{(b_f' - b)h_f'}{b h_0} = \frac{(890 - 310) \times 30.4}{310 \times 107} = 0.534$$

$$\rho_{te} = \frac{A_s}{0.5bh + (b_f - b)h_f} = \frac{28.3 \times 9}{0.5 \times 310 \times 120 + (890 - 310) \times 20.4} = 0.00838, \text{ take } \rho_{te} = 0.01.$$

$$\gamma_f' = \frac{(b_f' - b)h_f'}{b h_0} = \frac{(890 - 310) \times 20.4}{310 \times 102} = 0.374$$

$$\sigma_{sq} = \frac{M_q}{0.87 h_0 A_s} = \frac{2.91 \times 10^6}{0.87 \times 102 \times 28.3 \times 9} = 129\text{N/mm}^2$$

$$\psi = 1.1 - 0.65 \frac{f_{tk}}{\rho_{te} \sigma_{sq}} = 1.1 - 0.65 \times \frac{2.01}{0.01 \times 129} = 0.087 < 0.2, \text{take } \psi = 0.2.$$

$$B_s = \frac{E_s A_s h_0^2}{1.15\psi + 0.2 + \frac{6\alpha_E \rho}{1 + 3.5\gamma_f'}}$$

$$= \frac{2.1 \times 10^5 \times 28.3 \times 8 \times 102^2}{1.15 \times 0.2 + 0.2 + \frac{6 \times 0.056}{1 + 3.5 \times 0.374}} = 9.67 \times 10^{11} \text{N/mm}^2$$

$$B = \frac{M_k}{M_q(\theta - 1) + M_k} \cdot B_s = \frac{4.47}{2.91 \times (1.2 \times 2 - 1) + 4.47} \times 9.67 \times 10^{11} = 5.06 \times 10^{11} \text{N/mm}^2$$

$$f = \frac{5}{48} \times \frac{4.47 \times 10^6 \times 3040^2}{5.06 \times 10^{11}} = 8.5\text{mm} < \frac{l_0}{200} = \frac{3040}{200} = 15.2\text{mm}, \text{ the requirement is satisfied.}$$

8.2 Calculation for Crack Width in Normal Sections

8.2.1 Mechanism of Crack

Before cracking, the tensile stresses and strains in the steel and the concrete along the longitudinal direction of a member in service are generally uniformly distributed. When the tensile strain in the concrete approaches the ultimate tensile strain, all cross sections are going to crack. However, because the mechanical properties of the concrete are not homogeneous and local defects and micro-cracks are often induced by shrinkage and temperature change, the first (batch of) cracks for any test condition will occur in the weakest cross section, so the cracking position is random.

When the concrete in the outer edge of the zone reaches its tensile strength, it will not crack immediately due to the deformation of the concrete. When the strain is close to the limit strain value of concrete, it is in the state of the crack, as shown in Figure 8-6(a).

When the outside edge concrete reaches its ultimate tensile strain at the weakest cross section, one or several cracks will appear, as shown in Figure 8-6(b).

When the cracks appear, the sides of the crack are like a rubber band, but the retraction is not free which is restrained by the steel bar until it is blocked. In the retraction of the length l, because of the existance of the relative slip between concrete and reinforcing steel bar, defined as the bond stress τ^0, the tensile stress of concrete cracks has increased gradually with the increase of distance from the crack interface. After reaching the l, the bond stress vanishes, leading to the stress of the concrete more evenly distributed. As shown in Figure 8-6(b). In this case, l is the length of bond stress, or the length of transmission.

After the first crack appears within the bond stress function length l, the concrete ouside that part is still in a state of tension. Thus, when the bending moment continues to increase, new cracks may occur in another weak cross section, as shown in Figure 8-6(c).

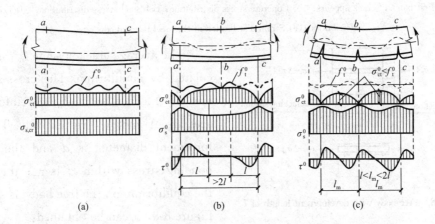

Figure 8-6 Crack propagation process
(a) Stress distribution before cracks; (b) The first batch of cracks; (c) Distribution and development of cracks

8.2.2 Average Crack Spacing

Obviously, the cross sections within l from the cracked sections and the cross sections within two adjacent cracks whose spacing is less than $2l$ cannot crack anymore, because the tensile stress in the concrete at those cross sections cannot be accumulated through the bond stress to the ultimate value. Therefore, theoretically, the minimum crack spacing is l, the maximum crack spacing is $2l$ and the average value is $l_m = 1.5l$.

With the increase of load, more and more cracks appear (section 2 in Figure 8-7). The variations of stresses and strains will redistribute until no more cracks appear. Consequently, the strains in the steel and the concrete are not uniform along the member.

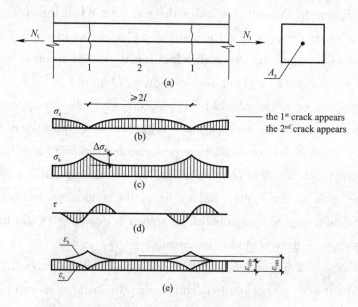

Figure 8-7　Crack propagation process

(a) The first batch of crack appears; (b) Concrete stress distribution; (c) Steel stress distribution; (d) Bond stress distribution; (e) Strain distribution of the steel bar and concrete

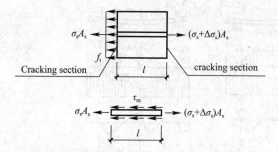

Figure 8-8　Free body with development length of l_{tr}

The average crack spacing l_m can be calculated by equilibrium. The cross-sectional area of the member and the reinforcement area are A and A_s, respectively. The reinforcement diameter is d and the average bonding stress within l_m is τ_m; then, from the equilibrium on the free body as shown in Figure 8-8, it can be obtained:

$$\Delta\sigma_s A_s = f_t A \tag{8-17}$$

And

$$\Delta\sigma_s A_s = \tau_m \pi d l \tag{8-18}$$

Thus

$$l = f_t A_s / \tau_m \pi d \tag{8-19}$$

$A_s = \pi d^2/4$, reinforcement ratio is $\rho = A_s/A$, and the average crack spacing is:

$$l_m = 1.5l = k_1' \frac{d}{\rho} \tag{8-20}$$

The value k_1' depends on τ_m and f_t. Experiments show that τ_m is proportional to f_t; therefore, their quotient is a constant, which means k_1' can be considered as a constant. According to the bond-slip theory, the average crack spacing is independent of the concrete strength. When the reinforcement type and the steel stress are constant, the variable to determine l_m is d/ρ which is proportional to l_m.

Experiments and analysis indicate that Eq. (8-20) also applies to flexural and eccentrically loaded members. Considering that the cross sections of flexural and eccentrically loaded members are not in full tension, the reinforcement ratio ρ can be substituted by the effective reinforcement ratio ρ_{te}, which is calculated by taking the sectional area of effectively tensioned concrete A_{te}. Then, Eq. (8-20) becomes

$$l_m = k_1 \frac{d}{\rho_{te}} \tag{8-21}$$

Where k_1——empirical coefficient.

The test also shows that l_m is not only related to d/ρ_{te}, but also has a great relationship with the thickness of concrete protective layer c. In addition, the average crack spacing of the ribbed deformed bar is smaller than that of the plain bar, indicating that the surface features of the bar also affect the average crack spacing and the equivalent diameter d_{eq} of steel bars can be used instead of d, so

$$l_m = k_2 c + k_1 \frac{d_{eq}}{\rho_{te}} \tag{8-22}$$

For the flexural members, the eccentric tension and the eccentric compression members can only use the expression of Eq. (8-22), but the values of the empirical coefficients k_2 and k_1 are different.

8.2.3 Average Crack Width

As mentioned above, the crack width refers to the crack width of the side surface at the level of the center of gravity of the section. Experiments show that the dispersion of crack width is larger than the crack spacing. Therefore, the average crack width should be based on the average crack spacing.

1. Formula of average crack width

The average crack width ω_m is equal to the difference between the average elongation of the rebar in the cracked section and the average elongation of the concrete on the side surface of the component at the corresponding level (Figure 8-9), so

$$\omega_m = \varepsilon_{sm} l_m - \varepsilon_{ctm} l_m = \varepsilon_{sm}\left(1 - \frac{\varepsilon_{ctm}}{\varepsilon_{sm}}\right) l_m \qquad (8\text{-}23)$$

Where ε_{sm} ——average tensile strain of longitudinally stretched bars, $\varepsilon_{sm} = \psi \varepsilon_{sq} = \psi \sigma_{sq}/E_s$;
  ε_{ctm}——the average tensile strain of concrete at the side surface at the same level of the longitudinal tensioned bar.

Defining

$$\alpha_c = 1 - \varepsilon_{ctm}/\varepsilon_{sm}$$

Where α_c——the impact coefficient of the concrete extension on the crack width, approximately $\alpha_c = 0.85$.

Thus,

$$\omega_m = \alpha_c \psi \frac{\sigma_{sq}}{E_s} l_m = 0.85 \psi \frac{\sigma_{sq}}{E_s} l_m \qquad (8\text{-}24)$$

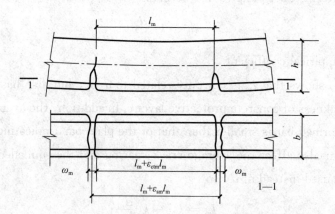

Figure 8-9 The average crack width calculation diagram

2. Reinforcement stress at crack section σ_{sq}

σ_{sq} refers to the stress of longitudinal tensile ordinary steel bars at the crack section calculated by quasi-permanent combination of loads. For the bending, axial tension, eccentric tension and eccentric compression members, σ_{sq} can be obtained according to the equilibrium condition of the crack section.

(1) Flexural members

σ_{sq} can be calculated as follows:

$$\sigma_{sq} = \frac{M_q}{0.87 A_s h_0} \qquad (8\text{-}25a)$$

(2) Axial tension member

$$\sigma_{sq} = \frac{N_q}{A_s} \qquad (8\text{-}25b)$$

Where N_q——axial force value calculated by permanent combination of loads;

A_s——the total cross-sectional areas of the tensile reinforcement.

(3) Eccentric tension member

The stress distribution in the fracture cross section of the large eccentric and small eccentric components are shown in Figure 8-10(a) and Figure 8-10(b).

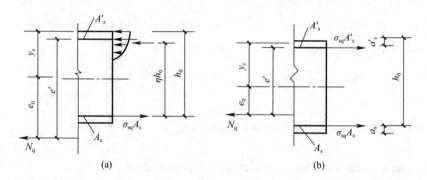

Figure 8-10 Stress calculation formula of steel bars for eccentric tension members

(a) Large eccentric tension; (b) Small eccentric tension

If the internal force arm length $\eta h_0 = h_0 - a'_s$ is approximately taken in the large eccentric tensile member, σ_{sq} can be unified by the following expression:

$$\sigma_{sq} = \frac{N_q e'}{A_s (h_0 - a'_s)} \tag{8-26}$$

Where e'——the distance between the point of axial tension force and the point of longitudinal reinforcement force in the area under compression or the area far from the axial force, $e' = e_0 + y_c - a'_s$;

y_c——the distance from the center of gravity to the edge far from the axial force.

(4) Eccentric compression member

The stress figure of the crack section under eccentrically compressed force is shown in Figure 8-11. Based on the moment of resultant force in compression zone, σ_{sq} can be obtained as:

$$\sigma_{sq} = \frac{N_q(e-z)}{A_s z} \tag{8-27}$$

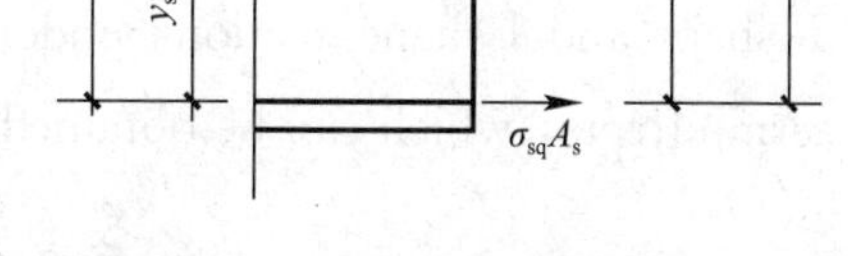

Where N_q——axial compression value calculated by quasi-permanent combination of loads;

e——the distance between the joint point of the tension bar to the point of the axial force, $e = \eta_s e_0 + y_s$, y_s is the distance from the center of gravity to the joint point of the tensioned steel bar in the longitudinal direction, η_s

Figure 8-11 Calculation formula of steel bar stress for eccentrically compressed members

refers to the axial compression eccentricity increase factor at the service stage, approximate as

$$\eta_s = 1 + \frac{1}{4000 e_0 / h_0} (l_0/h)^2 \tag{8-28}$$

If $l_0/h \leqslant 14$, $\eta_s = 1.0$ is taken.

z refers to the distance from longitudinal tensile steel reinforcement point to the resultant point of the compression zone; approximate as:

$$z = \left[0.87 - 0.12(1-\gamma'_f)\left(\frac{h_0}{e}\right)^2 \right] \tag{8-29}$$

8.2.4 Maximum Crack Width and Checking

1. The maximum crack width $\omega_{s,\,max}$ under the short-term load

It can be obtained by multiplying the average crack width by the crack width expansion factor τ, namely

$$\omega_{s,\,max} = \tau \omega_m$$

2. The maximum crack width ω_{max} under the long term load

Under long-term load, the concrete shrinkage will increase the crack width; at the same time as the concrete tensile stress relaxation and creep slip, the average strain of tensile steel on the crack section will continue to increase, leading to increase the width of the crack. The study shows that the maximum crack width under long-term load can be obtained by the maximum crack width under short-term load multiplied by the crack growth factor τ_l, namely

$$\omega_{max} = \tau_l \omega_{s,\,max} = \tau \tau_l \omega_m \tag{8-30}$$

Where $\tau_l = 1.5$; for axial tension member and eccentric tension members $\tau = 1.9$ and for eccentrically compression members $\tau = 1.66$.

Code for Design of Concrete Structures specifies the calculation method on tensile, bending and eccentric compression members of rectangular section, inverted T-shape, T-shape and I-shape sections under the quasi-permanent combination of loads and long-term effects, which can be obtained by the following equation:

$$w_{max} = \alpha_{cr} \psi \frac{\sigma_{sq}}{E_s} \left(1.9 c_s + 0.08 \frac{d_{eq}}{\rho_{te}} \right) (\text{mm}) \tag{8-31}$$

Where c_s——the distance between the outer edge of the outermost longitudinal tensile steel bar and the bottom edge of the tensile zone(mm): if $c_s < 20$mm, $c_s = 20$mm is taken; if $c_s > 65$mm, $c_s = 65$mm is taken;

σ_{sq}——the longitudinal reinforcement stress subjected by ordinary steel bars calculated in accordance with the quasi-permanent combination of loads;

d_{eq}——the equivalent diameter of the longitudinal tensile bar $d_{eq} = \sum n_i d_i^2 / \sum n_i v_i d_i$;

α_{cr}——the stress characteristics coefficient of members, for axial tension member, $\alpha_{cr}=2.7$; for eccentric tension member, $\alpha_{cr}=2.4$; for bending and eccentric compression members, $\alpha_{cr}=1.9$.

It should be noted that the maximum crack width calculated by Eq. (8-30) is not the absolute maximum value but the maximum crack width with a 95% guaranteed rate.

3. Checking on the maximum crack width

Code for Design of Concrete Structures GB 50010 divides the crack control levels of reinforced concrete members and prestressed concrete members into three grades. Grade 1 and 2 refer to prestressed concrete which requires no cracks. In the case of Grade 3 under control of cracks, the maximum crack width of reinforced concrete members can be calculated by permanently standard combinations of load considering long-term effects. The maximum width of cracks should meet the following requirements:

$$\omega_{max} \leqslant \omega_{lim} \qquad (8-32)$$

Where ω_{lim}——the maximum crack width limit specified in *Code for Design of Concrete Structures* GB 50010.

Same with the checking on deflection of flexural members, the calculation of crack width is also carried out under meeting requirements of the bearing capacity, when the cross-sectional dimension and the ratio of reinforcement have been determined. During checking process, the conditions that the deflection is met but the crack width not usually satisfies in the case of the members with lower reinforcement ratio and larger diameter of the selected steel. Therefore, when the maximum crack width exceeds not too much over the allowable value, the diameter of the reinforcement can be reduced; if necessary, the ratio of reinforcement may also be increased.

For bending and tension members, when high bearing capacity is required, the crack width or deflection limit requirements often cannot be met. At this time, increasing on the section size or the amount of steel are not effective. Instead, the prestress should be adopted.

In addition, attention should be paid to the relevant provisions in *Code for Design of Concrete Structures* GB 50010. For example, for the flexural member subjected to the crane load directly, the maximum crack width calculated can be multiplied by 0.85 due to the possibility of less loads and the value of $\psi=1$; for the eccentric compression members of $e_0/h_0 \leqslant 0.55$, the test shows that the maximum crack width is less than the allowable value, so there is no need for specific checking.

4. Limit value of maximum crack width

To determine the value limit of maximum crack width, two main considerations are involved, one is the appearance requirements; the other is durability requirements, which is the dominant.

Considering the appearance requirements, when the crack is too wide, people will get the impression of a sense of insecurity, and the evaluation of the quality is affected as well. The crack width limit satisfying the appearance requirement is related to people's psychological reaction, crack development length, crack location, and even light conditions. This aspect has not yet been studied, and currently there is a proposal for a desirable value of 0.25~0.3mm.

According to the investigation and test results domestic and abroad, the limit of crack width required by durability should take into account the environmental conditions and the working conditions of structural members. For the width of diagonal cracks, the width of cracks in the service stage is generally less than 0.2mm when there is sufficient shear resistance. Therefore, it is no need to check the width of the diagonal cracks.

[Example 8-2] The sectional size of a truss under the axial tension member is 200mm× 160mm with thickness of protective layer $c=25$mm. The configuration of longitudinal tension members are 4 ⌽ 16 of HRB400. The diameter of the stirrup is 6mm. The strength grade of concrete is C40. The axial tension under quasi-permanent combination of loads is $N_q=142$kN and $\omega_{lim}=0.2$mm. The maximum crack width is to be determined.

[Solution]

From Eq. (8-31), $\alpha_{cr}=2.7$

$$\rho_{te} = A_s/bh = 804/(200 \times 160) = 0.0251$$

$$d_{eq}/\rho_{te} = 16/0.0251 = 637 \text{mm}$$

$$\sigma_{sq} = N_q/A_s = 142000/804 = 177 \text{N/mm}^2$$

$$\psi = 1.1 - \frac{0.65 f_{tk}}{\rho_{te} \sigma_{sq}} = 1.1 - \frac{0.65 \times 2.39}{0.0251 \times 177} = 0.75$$

Thus,

$$\omega_{max} = \alpha_{cr} \psi \frac{\sigma_{sq}}{E_s} \left(1.9 c_s + 0.08 \frac{d_{eq}}{\rho_{te}}\right)$$

$$= 2.7 \times 0.75 \times \frac{177}{2 \times 10^5} [1.9 \times (25+6) + 0.08 \times 637]$$

$$= 0.197 \text{mm} < \omega_{lim} = 0.2 \text{mm}$$

8.3 Durability of Concrete Structures

Durability of a structure is defined as the ability of the structure to maintain its safety and serviceability when exposed to its intended service environment without spending too much money on repairing or rehabilitating.

Deterioration of durability for a structure is the degradation of structural performance arising from either intrinsic ingredients or external agents; thus, durability of a concrete structure depends on the structure and the service environmental conditions as well.

8.3.1 Carbonization

Concrete neutralization refers to the decreasing process of pH value, which is caused by the reaction between alkali in concrete and acid gas or liquid from air, soil or underground water. Because normal air contains carbon dioxide gas (CO_2), which can react with hydroxides to form carbonate. This CO_2 induced reaction, also referred to as carbonization, is the most common kind of concrete neutralization.

Carbonization is a complex physicochemical process. After sufficient hydration of cement, $Ca(OH)_2$ and $3CaO \cdot 2SiO_2 \cdot 3H_2O$ are produced. Pore solution tests show strong alkalinity with pH value of $12 \sim 13$ in the presence of saturated $Ca(OH)_2$. Stable water films come into being between pore water and surrounding moisture to achieve hydrothermal balance. With CO_2 continuously diffusing into concrete and dissolving in pore solution, solid $Ca(OH)_2$ dissolves and diffuses toward carbonated areas simultaneously by concentration gradient; thus, before the carbonation, CO_2 can react with $Ca(OH)_2$ and $3CaO \cdot 2SiO_2 \cdot 3H_2O$ in solution and solid-liquid interface, respectively.

In excessively high-moisture surroundings, carbonization nearly stops. This is becausethat CO_2 diffuses very slowly in concrete that is nearly in water saturated level. Under dry conditions, although CO_2 enters into concrete rapidly, the carbonization still develops very slowly without the necessary liquid phase. The most favorable relative humidity for concrete carbonization is between 70% and 80%.

8.3.2 Corrosion of Steel Embedded in Concrete

Corrosion of steel embedded in concrete is one of the key problems affecting the durability of concrete structures.

The damage on the oxide film on the surface of the steel bar is a necessary condition for the corrosion of the steel bar. At this time, if the oxygen-containing water intrudes, steel will rust. Therefore, the oxygen content of water intrusion is a sufficient condition for corrosion of steel. When the steel corrosion is serious, the volume expands, resulting in longitudinal cracks along the steel and the propective layer peeling. Therefore, the steel sections weaken, reducing the cross-section bearing capacity and leading to ultimate damage or failure of structural components.

The corrosion mechanism of steel bars in concrete is electrochemical corrosion. Due to the uneven distribution of chemical compositions in the steel, the difference in the alkalinity of the concrete and the concentration of oxygen at the crack, the potential difference between the parts of the steel surface is formed, which constitutes of a number of microcells with anodes and cathodes.

After the oxide film on the surface of the steel bar is damaged, the surface of the steel is contacting the air with water containing CO_2, O_2 or SO_2, and an electrolyte water film is formed in the micro-cell. As a result, an electrochemical corrosion reaction occurs between the cathode and the anode. The result is the formation of ferrous hydroxide, which in the air is further oxidized to ferric hydroxide, both rust of which are loose and porous with volume more than the original size of 2 to 4 times, leading to the cracks on the cover to further accelerate the development of corrosion.

The corrosion of steel undergoes a very long process, firstly it happens in the cracks at a wide point on the individual "pit corrosion", and then gradually forms to a "ring". When expanding to both sides as the formation of rust surface, the steel section weakens. When the corrosion is severe, the volume expands, resulting in longitudinal cracks along the length of the concrete and peeling off the concrete protective layer. The longitudinal crack along the steel bar can be used as a criterion to judge the end of life for concrete structural members.

The main measures to prevent corrosion of steels are:

(1) Reduction on the water-cement ratio, increase on the amount of cement and improvement on the density of concrete;

(2) Enough thickness of the concrete protective layer;

(3) Strict control of Cl^- content;

(4) The use of cover to prevent CO_2, O_2 and Cl^- infiltration.

8.3.3 Durability Design of Concrete Structures

Due to the complexity of the performance for the concrete structural materials, the uncertainty of the principles is very large. Generally, the durability design of the concrete structure can only be solved by empirical qualitative method. According to the provisions of the current national standard *Code for Durability Design of Concrete Structuves* GB/T 50476, the concrete design specification stipulates the basic contents of the durability design of concrete structures based on the research and the situation of China as follows:

1. The environmental category in which the structure is located is determined.

2. The basic requirement for the durability of concrete materials is put forward.

For the 50 years of design of the concrete structure, the basic requirements of the durability of concrete materials should be consistent with the provisions in Table 8-1 from GB 50010.

The basic requirement for the durability of concrete structure Table 8-1

Environment Classes	Maximum water cement ratio	Minimum strength level	MaximumCl⁻ Content (%)	Maximumalkali content(kg/m³)
Class 1	0.60	C20	0.30	not limited
Class 2-a	0.55	C25	0.20	3.0
Class 2-b	0.50(0.55)	C30(C25)	0.15	
Class 3-a	0.45(0.50)	C35(C30)	0.15	
Class 3-b	0.40	C40	0.10	

Note: (1) Cl⁻ content refers to the percentage of the total amount of cementitious materials; (2) The maximum content of Cl⁻ for the prestressed concrete is 0.06%; the minimum concrete strength grade should be increased by two levels according to the provisions in the table; (3) The water-cement ratio of the plain concrete components and the minimum strength level requirements may be appropriately reduced; (4) With reliable engineering experience, the lowest level of concrete strength in Class 2 can be reduced by a grade; (5) In the cold areas of Class 2-b and Class 3-a, concrete should add air entraining agent and can use the relevant parameters in parentheses; (6) For the use of non-alkali active aggregate, there are no restrictions of the amount of alkali in the concrete.

3. The thickness of the concrete protective layer of the reinforced concrete in the component is determined.

The thickness of the protective layer of the concrete should comply with the requirements of Table 4-3; when the effective surface protection is used, the thickness of the protective layer of the concrete may be appropriately reduced.

4. The durability measures should be used for concrete structures and components.

(1) Prestressed tendons in prestressed concrete structures should be taken according to the specific situation including the surface grouting, the thickness of the concrete protective layer and other measures. The exposed anchorage sections should take effective measures such as sealing or concrete surface treatment.

(2) For structures under impervious requirements, concrete impermeability grade should meet the requirements of the relevant standards.

(3) In cold and cold areas with humid environment, the structural concrete should meet the requirements of frost and concrete frost level should meet the requirements of the relevant standards.

(4) In the Class 2 and Class 3 environments, cantilever structures are suggested to use cantilever-plate structure or add protective layers in the upper surface.

(5) In the Class 2 and Class 3 environments of the embedded parts, hooks, connectors and other metal parts at the surface should take reliable anti-rust measures. For the ex-

posed metal anchor in the post-tensioned concrete, relevant requirements should be based on the chapter 10.3.13 in *Code for Design of Concrete Structures*.

(6) Concrete structures in Class 3 environment can adopt rust inhibitor, epoxy resin coated steel or other corrosion-resistant steel, the use of cathodic protection measures or replaceable components and other measures.

5. The detection and maintenance requirements of the structure in design life are proposed.

(1) The establishment for regular inspection and maintenance system;

(2) Regular replacement for the replaceable components in design;

(3) The protective layer on the surface of the component should be maintained in accordance with the regulations;

(4) The structure with visible defective defects should be promptly treated.

For the temporary concrete structure, no considerations are included in the durability requirements of concrete.

Questions

8-1 What is the bending stiffness of the section? What is the difference between the bending stiffness of the section and the stiffness in the material mechanics? How to calculate the stiffness of bending members?

8-2 What is the "minimum stiffness principle"? Try to analyze the rationality of applying this principle.

8-3 What is the effect of the concrete in the tension zone when checking the deflection and the crack width?

8-4 The physical meaning of parameter ψ, ρ_{te} is to be determined.

8-5 Briefly describe the emergence and distribution of cracks, the process and mechanism.

8-6 How is the maximum crack width calculation formula established? Why is the maximum value used rather than the average value?

8-7 Briefly introduce the influence of the reinforcement ratio on the bearing capacity, deflection and crack width of the normal section in the flexural members.

8-8 How is the calculation of the "combination of load standards considering the effect of quasi-permanent combination of loads" in the deflection and fracture width calculation formulas?

8-9 What is the ductility of the cross section of concrete? What is the main expression and influencing factor?

8-10 What are the main factors that affect the durability of concrete structures? What are the main contents of the durability design?

8-11 What factors are considered when determining the minimum thickness, component deformation and fracture limits for concrete coatings?

8-12 What is the axial compression ratio of the frame column? Why should the ratio meet the requirements of the axial compression ratio limit?

Exercises

8-1 The reinforced concrete bottom chord of a roof truss with a section of $b \times h = 200\text{mm} \times 200\text{mm}$ is subjected to an axial tension under the permanently combined of load effect $N_q = 130\text{kN}$. The steel bars are 4 ⏀ 14 and concrete strength grade is of C30 with the protective layer thickness $c = 20\text{mm}$. The stirrup diameter is 6mm and the crack width limit is $\omega_{\lim} = 0.2\text{mm}$. Is the crack width satisfied? If not, what measures should be taken?

8-2 Figure 8-12 shows the rectangular section of a simply supported beam with the span $l = 6\text{m}$. The concrete is of grade C30 and the steel is of HRB500. The beam is subjected to uniformly distributed loads. In mid-span, the moment under quasi-permanent combination load effect is $M_q = 100\text{kN} \cdot \text{m}$. It is required to check the deflection of the beam.

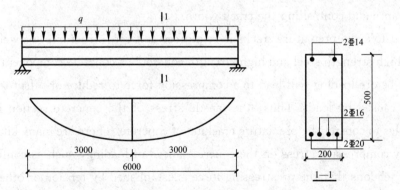

Figure 8-12 Exercise 8-2 (Unit: mm)

Chapter 9
Prestressed Concrete Members

9.1 Introduction

9.1.1 Basic Concept of Prestressed Concrete

With the low tensile strength and low ultimate tensile strain, reinforced concrete structures usually work with cracks. In order to satisfy the requirement of deformation and crack, it is necessary to increase the section size of the component and the amount of steel. However, that will lead to too much weight, which is uneconomic and impossible in large span or under dynamic load. If high strength steel is taken, the stress usually can reach $500 \sim 1000 \text{N/mm}^2$ under service, which cannot meet the requirements of the crack width. Therefore, high strength steel bars in the reinforced concrete structure cannot fully be taken use of. And it is essential to improve the concrete strength grade for improving the crack resistance and controlling the crack width.

In order to avoid premature cracks in reinforced concrete structures, as well as making full use of high strength steel and high strength concrete, a concrete component that can be prestressed before loading will lead to a compression force to reduce or offset tensile stress of concrete caused by loads. Thus, the tensile stress of the concrete section is not large which enables to control the premature cracking of concrete. There are many kinds of methods to apply compressive stress on the concrete ahead of loading, such as configuration of prestressed tendons that the prestressing force is established by tension or other methods, or self-stressing concrete produced by expansive concrete in the centrifugal pipe and so on. It refers to the prestressed concrete members with tensioning prestressing tendons commonly used in this chapter.

The concept of prestressed concrete can be illustrated by the simply supported beam shown in Figure 9-1.

The beam is prestressed by exerting an eccentric compression N on the tensile side of the beam before the beam is subjected to any external load. This will cause the beam section subjected to prestressed compression σ_c at the bottom fiber and prestressed tension σ_{ct} at the top fiber (Figure 9-1a). Under the action of the external load q(including the weight of the beam), tensile stress is defined as σ_{ct} at the bottom fiber of the middle section and compressive stress is defined as σ_c at the top fiber (Fig 9-1b). In this way, under the combined action of prestressed compression N and the load q, the tensile stress at the bottom fiber of the beam will be reduced to $\sigma_{ct} - \sigma_c$ and the stress at the top fiber is equal to $\sigma_c - \sigma_{ct}$ with tension or compression (Figure 9-1c). If the compressive prestress N is large enough, the tensile stress at the bottom fiber of the beam under the action of the load q can even turn to the compressive stress. It can be seen that the prestressed concrete members can delay the cracking of concrete components and improve the crack resistance and stiffness of the components.

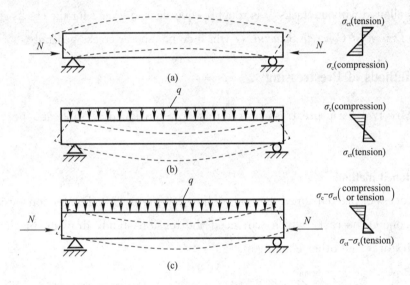

Figure 9-1 Simply supported prestressed beam

With many advantages over reinforced concrete, prestressed concrete should be applied in priority under such conditions:

(1) Structures with strict crack-control;

(2) Components that are used for large-span or subjected to huge force;

(3) Structural members that require a higher level of stiffness and deformation control, such as crane beams of the industrial plant and large-span beam components of the piers and bridges.

Compared with ordinary reinforced concrete structures, prestressed concrete structures have some disadvantages. For example, it is complicated for structure, as well as construction and calculation, and the ductility is a little poor.

9.1.2 Classification of Prestressed Concrete

Based on the differences of the degrees of prestressing, prestressed concrete members can be divided into full prestressed members and partial prestressed members.

Full prestressed concrete is the member that does not allow the tensile stress in the concrete section under service loads, roughly equivalent to level 1 for the cracks control level of *Code for Design of Concrete Structures*, which is strict in cracking.

Partially prestressed concrete includes members in which the cracks are allowed to appear according to the load combination under service loads with the crack width not exceeding the corresponding allowable value. It is roughly equivalent to level 3 for the cracks control level of *Code for Design of Concrete Structures*, which is allowable in crack.

Limit value prestressed concrete belongs to partially prestressed concrete, in which the tensile stress is allowed without cracks. It is roughly equivalent to level 2 for the cracks control level of *Code for Design of Concrete Structures*, which do not appear cracks generally.

9.1.3 Methods of Prestressing

There are two basic prestress procedures: pre-tensioned method and post-tensioned method.

1. Pre-tensioned method

The process of pre-tensioned method is shown in Figure 9-2. The tendons are tensioned before the concrete is cast, which will need the tension stand, drawing machine, dowel frame and fixture, and other equipment.

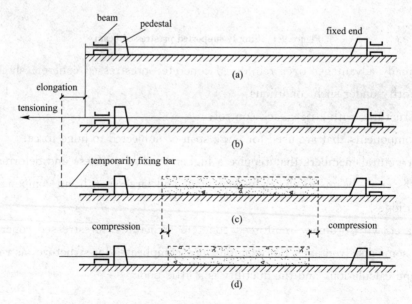

Figure 9-2　The main process of pretensioning

(a) Tendon in place; (b) Tensioning tendon; (c) Temporarily fixing bar, pouring concrete and curing;
(d) Relaxing tendon with retraction, leading to concrete under pre-compression

Under this method, the transfer of prestress is mainly by the bond between the tendon and concrete.

2. Post-tensioned method

The process of posttensioning is as follows (Figure 9-3). Firstly, the member is cast with ducts. Then, it is cured. When the concrete of the member reaches sufficient strength, the tendons are put through the ducts.

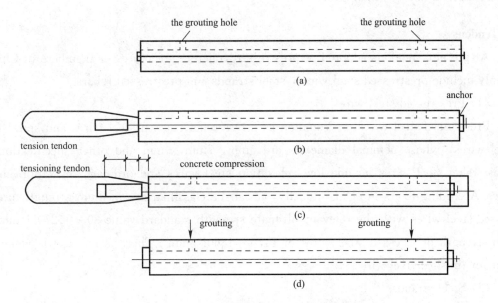

Figure 9-3　The main process of posttensioning

(a) Casting members with free tendons in the ducts; (b) Installing jack; (c) Tensioning tendon;
(d) Anchoring tendon with disassembling the jack and grouting hole

Obviously, in this case, the prestress is transferred to the concrete by the anchors at the two ends of the tendon.

9.1.4　Materials of Prestressed Concrete

1. Concrete

The concrete used in prestressed concrete structures should satisfy the following requirements.

(1) High strength. Different from reinforced concrete, the prestressed concrete should use high strength concrete. For pretension members, the high-strength concrete can increase the bond strength between the tendon and the concrete. For posttension members, the high-strength concrete enables the anchorage region to withstand higher local compression.

(2) Less shrinkage and creep. This behavior can reduce the prestress loss due to shrinkage and creep.

(3) Rapid hardening and higher early-age strength. This performance enables the prestressing at an early age, resulting in more efficient construction.

Therefore, *Code for Design of Concrete Structures* specifies that the grade of concrete used in prestressed concrete structures should not be less than C30 with a preference for a concrete grade not less than C40.

2. Tendons

At present, tendons adopted in prestressed concrete structures or members in China mainly include prestressed steel wires, steel strands and prestressed rebars.

(1) Prestressed steel wires

Prestressed steel wires are common as stress-relief smooth steel wires and spiral rib steel wires, whose nominal diameters are 5mm, 7mm, 9mm and other specifications. Stress-relief steel wires include low relaxation steel wires and ordinary relaxation steel wires. According to its level of intensity, it can be classified as: medium strength prestressed steel wires with the relevant ultimate strength standard value $800 \sim 1270 N/mm^2$; high strength prestressed steel wires of $1470 \sim 1860 N/mm^2$. There is no electric welding joint for finished steel wires.

(2) Steel strands

Steel strands are made by twisting several steel wires together, such as a 2-wire, 3-wire or 7-wire strand. Steel strands commonly twisted by three wires are expressed as 1×3, of which the nominal diameter is $8.6 \sim 12.9mm$. Standard steel strands twisted by seven wires are expressed as 1×7, of which the nominal diameter is $9.5 \sim 21.6mm$. Prestressed tendons are often composed of a plurality of steel strands. For example, the specifications for prestressed steel strands are 15-7Φ9.5, 12-7Φ9.5, 9-7Φ9.5 and so on. The characteristic tensile strength of a steel strand can be as high as $1960N/mm^2$. And the resistance relaxation property of steel wire is good, keeping upright when expanded. Steel wires should not be broken with transverse cracks or intersected and the surface must be free from oil, grease and other substances, so as to ensure the bond strength between steel and concrete. The surface of steel strands can have slight floating rust, but there cannot be visual corrosion pits.

(3) Prestressed rebars

The rebars for prestressed concrete are made into external-thread straight reinforcement with discontinuous and no longitudinal ribs by process of hot rolling, remained heat treatment after rolling or heat treatment and so on. The bars can be connected or anchored by matching the shape of internal thread connectors or anchorages at any cross section. The diameter ranges from 18mm to 50mm with the characteristics of high strength, high toughness and so on. It requires that the steel end should be flat, in case of the failure for the connector passing through. The surface should not have transverse cracks or scarring, but in allowance with some defects showing no influence on the mechanical properties of the steel and connection.

9.1.5 Tension Control Stress σ_{con}

The tension control stress is the permissible maximum stress in a prestressed tendon during tensioning. The value is measured as the stress obtained by the total tension indicated from the tension device divided by the area of the prestressed tendon.

The choice of the tension control stress value will directly affect properties of the prestressed concrete. If the tension control stress is too low, the prestress on concrete will also be very low after various losses of prestress, which will not effectively increase the resistance to cracking and stiffness of the members. However, it may cause the following problems if the control stress is too high:

(1) The tensile stress (usually in pretension) by the prestressed method may lead to cracks in some parts of the component during the construction stage. In the case of posttensioning, high control stress may cause the end of the member subjected to a local bearing failure.

(2) The cracking load may be too close to the ultimate load with little warning before failure, of which the ductility in the member may be poor.

(3) In order to reduce the prestress loss, over tensioning is sometimes applied. It is possible that the stress of individual prestressed tendon exceeds the actual yield strength in the process of over tensioning, which may cause large plastic deformation or brittle fracture.

The choice of tension control stress is also related to types of prestressed reinforcement. Because of high strength steel bar in the prestressed concrete with poor plastic performance, the control stress should not be too high.

The allowable tension control stress, as specified by *Code for Design of Concrete Structures* is given in Table 9-1. These values are generally not to be exceeded.

The limit of the tension control stress in Table 9-1 can be increased $0.05 f_{ptk}$ or $0.05 f_{pyk}$ in one case of the following circumstances:

(1) It is required to improve the crack-resistance performance of the component during the construction stage and the prestressed tendons are installed in the compression zone un-

der service;

(2) It is required to partially offset prestress loss caused by stress relaxation, friction, batch tensioning, the temperature difference between the stretching bed and prestressed tendons and other prestress losses.

Allowable tension control stress [σ_{con}]　　　　Table 9-1

Steel type	σ_{con}
Stress relieving steel wire, steel strand	$\leqslant 0.75 f_{ptk}$
Medium strength prestressed steel wire	$\leqslant 0.70 f_{ptk}$
Prestressed rebar	$\leqslant 0.85 f_{pyk}$

Note: (1) The tension control stress value should not be less than $0.4 f_{ptk}$ in stress relieving steel wire, steel strand and medium strength prestressed steel wire in table;
(2) The tension control stress value should not be less than $0.5 f_{pyk}$ in prestressed rebar in table.

9.1.6　Prestress Losses

The prestress established by jacking will gradually become less in the process of construction and service due to the nature of construction and material properties. This reduction in tendon stress after jacking is called prestress loss. Many factors will cause prestress losses. It is generally accepted that the total prestress loss is equal to the summation of the individual prestress losses, which can be calculated by the method of superposition through the prestress loss values from various factors. The six different types of prestress losses are given below.

1. Prestress loss in linear prestressed reinforcement due to anchorage deformation and prestressed tendons retraction σ_{l1}

After the linear prestressed reinforcement is tensioned to σ_{con}, anchored on the pedestal or member, the tendon will shorten due to the relative displacement and local plastic deformation between the anchorage parts (such as the gap between the anchor, the backing plate and member under compression) and between the prestressed tendon and anchorage. Hence, the prestress loss σ_{l1} can be expressed by the following formula for straight tendons:

$$\sigma_{l1} = \frac{a}{l} E_s \qquad (9\text{-}1)$$

Where　a——the deformation of tensioning end anchorage and the internal shrinkage value of prestressed tendons (mm), as obtained from Table 9-2;

　　　　l——distance between the anchored end and the jacked end (mm);

　　　　E_s——modulus of elasticity of the tendon (N/mm²), as obtained from Table 1-14 in Appendix 1.

Anchorage deformation and internal shrinkage value of prestressed tendons α(mm)   Table 9-2

Types of anchorages		α
Bearing-type anchorages (anchorages for wire bundle with extended ends)	Gap due to the nut	1
	Gap from each extra gasket	1
Clamping-type anchorages	With pressing pressure	5
	Without pressing pressure	6~8

Note: (1) The values of shortening in the table can also be determined by measurement;
(2) The values of shortening for other types of anchorages not mentioned should be determined by measurement.

The prestress loss due to anchorage deformation is considered only for the tensioned end. There is no such loss at the anchored end, because the anchorage at this end has already been pressed tightly during the jacking process.

For structures assembled with several segments, the prestress losses should include the deformation under pre-compression caused by gaps between the segments. Such losses can be assumed as 1 mm shortening for each gap if the gap is filled by concrete or mortar.

The measures to reduce σ_{l1} are as follows:

(1) Anchorages and clamps should be carefully selected with less deformations or less slides. As few gaskets as possible should be used. For each extra gasket, α is increased by 1mm.

(2) The length of the platform should be increased. σ_{l1} is inversely proportional to the length of the platform. Therefore when the length of the platform is 100m or longer in pretensioning production, σ_{l1} can be omitted.

For post tensioned member with curved prestressed tendons or polyline prestressed tendons, due to deformation of anchorage and shrinkage of prestressed tendons, the prestress loss, σ_{l1} should be determined by the following condition: the deformation value of prestressed tendon in the range of reverse friction impact length l_f between the curved prestressed tendons or polyline tendons and pipe wall, is equal to the anchorage deformation and prestressed tendons in shrinkage. σ_{l1} can be calculated according to appendix J of *Code for Design of Concrete Structures*.

2. Prestress loss σ_{l2} due to friction between tendon and duct

In the case of posttensioning with a prestressed tendon, friction will be generated due to the contact between tendons and concrete hole wall or bushing when tensioning. The effect of such friction will be greater for the position with greater distance from the jacked end. The actual prestress on each section of the component is reduced (Figure 9-4). The prestress loss due to such friction is denoted as σ_{l2}, which can be calculated as the following method.

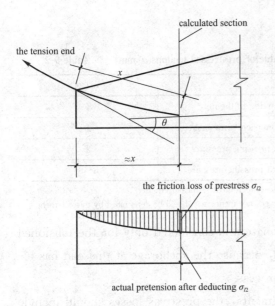

Figure 9-4 Prestress loss caused by friction

The friction is mainly caused by the following two reasons, which should be firstly calculated separately and then added together:

(1) As tensioning curved prestressed tendons, due to the curvature of the curved channel, the friction resistance is caused by the normal pressure between the prestressed tendon and the hole wall (Figure 9-5b).

The tensions at both ends of the dx section are referred to as N and $N\text{-}dN'$, and the pretensions of the dx at the both ends to the hole wall are

$$F = N\sin\left(\frac{1}{2}d\theta\right) + (N - dN')\sin\left(\frac{1}{2}d\theta\right)$$
$$= 2N\sin\left(\frac{1}{2}d\theta\right) - dN'\sin\left(\frac{1}{2}d\theta\right)$$

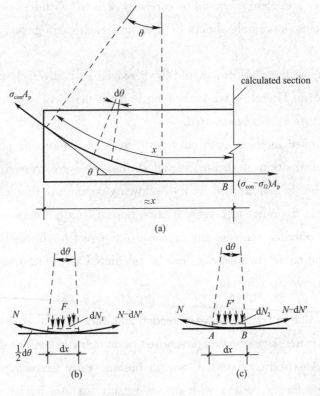

Figure 9-5 Friction between tensioned steels and the hole wall in reserved hole

$\sin\left(\frac{1}{2}d\theta\right) \approx \frac{1}{2}d\theta$ is taken and $dN'\sin\left(\frac{1}{2}d\theta\right)$ is ignored due to the smaller values:

$$F \approx 2N\frac{1}{2}d\theta = Nd\theta$$

Supposed that the friction coefficient between the prestressed tendon and the channel is μ, the frictional resistance of the dx section dN_1, is

$$dN_1 = -\mu N d\theta$$

(2) Due to some reasons of the reserved channel at the stage of construction, such as local deviation, wall roughness and prestressed tendons deviating from the designed position, normal pressure may be generated between tendons and pore wall on tensioning the prestressed tendons, which will cause friction resistance.

The degree of discrepancy between the channel position and the design position is indicated by deviation coefficient k', which is the deviation value of unit length. The angle that B end deviates from A end is taken as $k'dx$, so the normal pressure on the dx segment exerted by the tendon on the hole wall can be derived as

$$F' = N\sin\left(\frac{1}{2}k'dx\right) + (N - dN')\sin\left(\frac{1}{2}k'dx\right) \approx Nk'dx$$

The frictional force dN_2 acting on the dx segment is

$$dN_2 = -\mu N k' dx$$

The above two frictional resistance dN_1 and dN_2 are summed up, followed by the integral equation from the tension end to the calculation section point B, thus:

$$dN = dN_1 + dN_2 = -[\mu N d\theta + \mu N k' dx]$$

$$\int_{N_0}^{N_B}\frac{dN}{N} = -\mu\int_0^\theta d\theta - \mu k'\int_0^x dx$$

Where μ, k' are both experimental values, and $\mu k'$ is taken by friction coefficient of per-meter local deviation of the channel, which is

$$\ln\frac{N_B}{N_0} = -(kx + \mu\theta)$$

$$N_B = N_0 e^{-(kx+\mu\theta)}$$

Where N_0——tension of the jacking end;

N_B——tension at the point B.

Now, the tension loss N_{l2} from the tension end to the point B can be derived as

$$N_{l2} = N_0 - N_B = N_0[1 - e^{-(kx+\mu\theta)}]$$

Divided by the section area of the prestressed tendon, we have:

$$\sigma_{l2} = \sigma_{con}[1 - e^{-(kx+\mu\theta)}] = \sigma_{con}\left(1 - \frac{1}{e^{kx+\mu\theta}}\right) \tag{9-2}$$

When $(kx + \mu\theta) \leqslant 0.3$, σ_{l2} can be approximately expressed as

$$\sigma_{l2} = (kx + \mu\theta)\sigma_{con}$$

To be mentioned, when the clamping system is used, the friction loss of anchor mouth should be deducted in σ_{con}, which is determined according to the measured value or the manufacturer.

Where k——friction coefficient of per-meter local deviation of the channel, according to the Table 9-3;

x——distance in meters between the calculated section and the jacked end, which can be adopted as approximately the projection length on the longitudinal axis of the channel (Figure 9-5);

μ——friction coefficient between prestressed tendon and channel wall, obtained from Table 9-3;

θ——angle in radians between the tangent of the tendon at the jacked end and the tangent of the tendon at the calculated section.

Coefficients of friction Table 9-3

Type of duct	k	μ	
		Steel wires and steel strands	Prestressed rebar
Pre-embedded corrugated metal tube	0.0015	0.25	0.5
Pre-embedded corrugated plastic tube	0.0015	0.15	—
Pre-embedded steel tube	0.0010	0.30	—
Formed by core withdrawal	0.0014	0.55	0.60
Unbonded tendon	0.0040	0.09	—

Note: The coefficient of friction can also be determined according to the measured data.

In Eq. (9-2), for the space curve of variation along the parabola or circular arc and the generalized space curve of the piecewise superposition, the angle θ can be calculated as follows:

Parabola, arc curve: $\theta = \sqrt{\alpha_v^2 + \alpha_h^2}$

Generalized space curve: $\theta = \sum \sqrt{\Delta\alpha_v^2 + \Delta\alpha_h^2}$

Where α_v, α_h——bending angle that is generated by projection in the vertical and horizontal direction from space curved prestressed tendons of variation along the parabola or circular arc;

$\Delta\alpha_v$, $\Delta\alpha_h$——bending angle increment of the piecewise curve formed by projection in the vertical and horizontal direction from the generalized space curved prestressed tendons.

Measures to reduce σ_{l2} include:

(1) Jacking at both ends of the tendon can be performed for long members. In this

case, the duct length used in calculation can be reduced by half. The effect of this measure in reducing the friction loss is obvious as shown in Figure 9-6(a) and Figure 9-6(b). However, this measure will increase σ_{l1}.

(2) Over tensioning can be used as shown in Figure 9-6(c). The process of over tensioning can be:

$$1.1\sigma_{con} \xrightarrow{\text{stop-for-2min}} 0.85\sigma_{con} \xrightarrow{\text{stop-for-2min}} \sigma_{con}$$

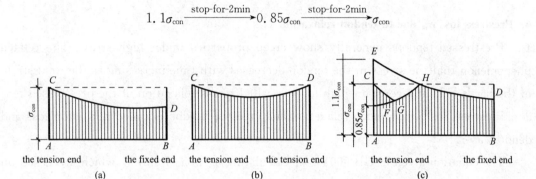

Figure 9-6 The effect of reducing friction loss by tensioning in the one end, two ends and over tensioning
(a)Tensioning in one end; (b)Tensioning in two ends; (c)Over tensioning

When the end A is over tensioned by 10%, the distribution of tendon stress is represented by curve EHD in the figure. When the tensioning stress is lowered to $0.85\sigma_{con}$, the distribution of tendon stress is given by curve FGHD as the tendon is now affected by reversed friction. When the tensioning end A finally reaches σ_{con} again, the distribution is given by curve CGHD which is obviously more uniform with less prestress loss.

3. **Prestress loss σ_{l3} caused by temperature difference between prestressed tendon and equipment under the tension of concrete heating**

Steam curing is often used to reduce the time spent on the production of pretensioned members. When the temperature is raised, the tendon can expand freely resulting in prestress loss.

The temperature difference between the tendon and the abutment platform is taken as $\Delta t(\text{℃})$ as the member is being steam cured. If the expansion coefficient of the tendon is $0.00001/\text{℃}$, then σ_{l3} can be derived as

$$\sigma_{l3} = \varepsilon_s E_s = \frac{\Delta l}{l} E_s = \frac{\alpha l \Delta t}{l} E_s = \alpha E_s \Delta t$$
$$= 0.00001 \times 2.0 \times 10^5 \times \Delta t = 2\Delta t (\text{N/mm}^2) \tag{9-3}$$

Measures to reduce σ_{l3} include:

(1) The process of cure in which the temperature is raised twice can be adopted. That is, the room temperature is maintained until the concrete reaches certain strength, e. g., 7.5~10MPa, then the temperature is further raised to a higher curing temperature at

which the concrete and the tendon have bonded together allowing no prestress loss due to temperature difference;

(2) The tendon can be tensioned in a steel form which is also subjected to the curing steam. Thus, there is no temperature difference between the tendon and the form. Hence, no prestress loss will be caused due to this process.

4. Prestress loss σ_{l4} due to tendon relaxation

Prestressed tendons generally show creep properties under high stress. The related phenomenon that the stress in the tendon decreases with time increasing as the total strain of the bar is held at a constant high level is called stress relaxation of the prestressed tendon. Prestress loss due to such relaxation and creep of tendon is called relaxation loss and denoted as σ_{l4}.

The equation given in GB 50010 for calculating σ_{l4} is as follows, which are based on test results.

(1) Stress-relief steel wires and steel strands

1) Ordinary relaxation:

$$\sigma_{l4} = 0.4\left(\frac{\sigma_{con}}{f_{ptk}} - 0.5\right)\sigma_{con} \tag{9-4}$$

2) Low relaxation:
When $\sigma_{con} \leqslant 0.7 f_{ptk}$,

$$\sigma_{l4} = 0.125\left(\frac{\sigma_{con}}{f_{ptk}} - 0.5\right)\sigma_{con} \tag{9-5}$$

When $0.7 f_{ptk} < \sigma_{con} \leqslant 0.8 f_{ptk}$,

$$\sigma_{l4} = 0.2\left(\frac{\sigma_{con}}{f_{ptk}} - 0.575\right)\sigma_{con} \tag{9-6}$$

(2) Medium-strength prestressed steel wire:

$$\sigma_{l4} = 0.08\sigma_{con} \tag{9-7}$$

(3) Prestressed rebar:

$$\sigma_{l4} = 0.03\sigma_{con} \tag{9-8}$$

When $\sigma_{con}/f_{ptk} \leqslant 0.5$, $\sigma_{l4} = 0$.

According to experiments, the following factors influence the degree of relaxation loss:

(1) Time: the development of relaxation loss is faster at the initial stage such that the relaxation loss within the first one hour can be as high as 50% of the total relaxation loss and 80% after the first hour. The development of relaxation loss will be slower afterward.

(2) Initial stress and ultimate strength of steel: when the initial stress is lower than $0.7 f_{ptk}$, relaxation is linear with initial stress; when the initial stress is higher than 0.7

f_{ptk}, the relaxation is significantly increased.

(3) The higher the control stress is, the greater the amount of relaxation loss will be; on the contrary, it is the opposite tendency.

The measure to reduce σ_{l4} is over tensioning. The process for this over tensioning is as follows. Firstly, the tensioning stress is raised to $(1.05 \sim 1.1)\sigma_{con}$, kept at this level for $2 \sim 5$ mins. Then, the stress is released. After that the stress is finally raised to σ_{con}. The reason for this process to reduce the relaxation loss is that the relaxation caused by short-term high stress can be as large as the relaxation caused by long-term low stress. Therefore, most of the relaxation loss has taken place in the over tensioning process.

5. **The loss value of the longitudinal prestressed tendons in the tension zone and compression zone caused by the shrinkage and creep of concrete σ_{l5} and σ'_{l5}.**

Concrete will shrink when hardened in air under normal temperatures. The action of prestress will cause the concrete creep in the direction of compression, which will lead to the member and the tendon shortening, resulting in prestress loss. Although creep and shrinkage are two different phenomena, their effects for prestress loss are usually taken into account as a whole due to their similar influencing factors and similar rules of variation.

The prestress loss due to creep and shrinkage of the tendon in the tension zone is denoted as σ_{l5} and in the compression zone is denoted as σ'_{l5}. They can be calculated as follows.

For pretensioned members

$$\sigma_{l5} = \frac{60 + 340 \frac{\sigma_{pc}}{f'_{cu}}}{1 + 15\rho} \tag{9-9}$$

$$\sigma'_{l5} = \frac{60 + 340 \frac{\sigma'_{pc}}{f'_{cu}}}{1 + 15\rho'} \tag{9-10}$$

For posttensioned members

$$\sigma_{l5} = \frac{55 + 300 \frac{\sigma_{pc}}{f'_{cu}}}{1 + 15\rho} \tag{9-11}$$

$$\sigma'_{l5} = \frac{55 + 300 \frac{\sigma'_{pc}}{f'_{cu}}}{1 + 15\rho'} \tag{9-12}$$

Where σ_{pc}, σ'_{pc}——the normal compressive stresses in concrete at the resultant points of the tendons in the tension and compression zone, respectively. In this case, only the prestress loss before pre-compression (first batch loss) is considered, and the corresponding σ_{l5} and σ'_{l5} in the ordinary rebar should be taken as zero. The values of σ_{pc} and σ'_{pc} should not exceed 0.5

f'_{cu}. Furthermore, if σ'_{pc} is tensile, the values of σ'_{pc} in Eqs. (9-10) and (9-12) should be taken as zero. In the calculation of σ_{pc} and σ'_{pc}, the self-weight of the member should be considered according to the actual condition during the process of construction;

f'_{cu}——the cube compressive strength of concrete when the prestress is applied;

ρ, ρ'——the reinforcement ratios of the total steel bars including both prestressed and ordinary steel bars in the tension and compression zones, respectively.

For pretensioned members

$$\rho = \frac{A_p + A_s}{A_0}, \quad \rho' = \frac{A'_p + A'_s}{A_0} \tag{9-13}$$

For posttensioned members

$$\rho = \frac{A_p + A_s}{A_n}, \quad \rho' = \frac{A'_p + A'_s}{A_n} \tag{9-14}$$

Where A_0——the area of the transformed section;

A_n——the net area of the concrete section.

For members with symmetric reinforcement of both tendons and steel rebars, the reinforcement ratios ρ and ρ' are taken as half of the total steel area (Figure 9-7).

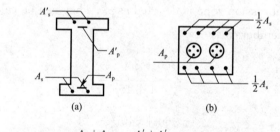

Pretensioning: $\rho = \frac{A_p + A_s}{A_0}, \rho' = \frac{A'_p + A'_p}{A_0}$ Pretensioning: $\rho = \rho' = \frac{A_p + A_s}{2A_0}$

Posttensioning: $\rho = \frac{A_p + A_s}{A_n}, \rho' = \frac{A'_p + A'_s}{A_n}$ Posttensioning: $\rho = \rho' = \frac{A_p + A_s}{2A_n}$

Figure 9-7 Reinforcement ratios ρ and ρ' when calculating σ_{l5}

(a) Flexure member; (b) Axial tension member

Eqs. (9-9)~(9-12) indicate the following:

1) The relationship between σ_{l5} and σ_{pc}/f'_{cu} is linear, the equation is given for the stress loss under the condition of linear creep, so it is required to meet the requirement of $\sigma_{pc} < 0.5 f'_{cu}$. Otherwise, the prestress loss will increase out of proportion.

2) The σ_{l5} of a posttensioned member is lower than that of a pretensioned member, for some of the shrinkage loss of the former has already taken place when the member is prestressed by tensioning.

In the case that the environment is so dry that the annual average humidity is less than 40%, the σ_{l5} and σ'_{l5} values should be increased by 30%.

Measures for reducing σ_{l5} include:

1) High-grade cement, reduction in cement content, reduction in water cement ratio and dry hardening concrete are suggested;

2) Aggregates with well grading and fully vibration to the fresh concrete to increase the compactness of the concrete are suggested;

3) Careful curing on the concrete is to reduce shrinkage of concrete.

6. Prestress loss σ_{l6} due to local compression of concrete for annular member with helical tendons

For annular members with helical tendons, the tendon will press the concrete leading to reduction in the diameter of the member. This will reduce the tension in prestressed tendon resulting in prestress loss, which is denoted as σ_{l6}.

The value of σ_{l6} is in reverse proportion to the diameter d of the member. The simplified calculation of σ_{l6} is as follows:

$$\text{When } d \leqslant 3\text{m}, \ \sigma_{l6} = 30\text{N/mm}^2 \qquad (9\text{-}15)$$

$$d > 3\text{m}, \ \sigma_{l6} = 0 \qquad (9\text{-}16)$$

9.1.7 Combination of Prestress Losses

The above prestress losses should be combined into batches to facilitate the calculation. The method of combination given in the code GB 50010 is shown in Table 9-4.

Combination of prestress losses at each stage   Table 9-4

Combination of prestress losses	Pretensioned members	Posttensioned members
Losses before prestress transfer (first batch)$\sigma_{l\text{I}}$	$\sigma_{l1}+\sigma_{l2}+\sigma_{l3}+\sigma_{l4}$	$\sigma_{l1}+\sigma_{l2}$
Losses after prestress transfer (second batch)$\sigma_{l\text{II}}$	σ_{l5}	$\sigma_{l4}+\sigma_{l5}+\sigma_{l6}$

Note: The according distribution of σ_{l4} between the first batch and the second batch can be determined to real conditions when needed.

Taking into account the discrete nature of the prestress loss, the actual prestress loss could be higher than the calculated value. Therefore, if the calculated prestress loss is less than the following values, these larger values should be adopted instead.

Pretensioning members: 100N/mm²

Posttensioning members: 80N/mm²

When prestressed tendons for posttensioning member adopt different batches of tension, the influence on the first batch of prestressed tendon should be considered, which is caused by the elastic compression (or elongation) of concrete reduced by the post batch of tensioning prestressed tendon. The first batch of tensioning control stress value σ_{con} should increase (or decrease) $\alpha_E \sigma_{pci}$, where σ_{pci} is the normal stress of concrete at the center of gravity of the first batch prestressed tendon which is caused by the post batch of tensioning prestressed tendon.

9.1.8 Transfer Length of Pretensioned Tendons

It is through the bond between the tendon and the concrete that the force in the tendon is transferred to the concrete during pretensioning. This transfer is within a short distance at the end of the member.

The segment of the tendon within a distance x from the member end is taken out as a free body (Figure 9-8). At the time of prestress transfer, the tendon will shorten and tend to slip with respect to the concrete. At the end point a, the prestress of the tendon is zero. Within the member, the shortening of the tendon is resisted by the concrete, resulting in tension stresses σ_p in tendon and compressive stresses σ_c in the concrete. With increasing value of x, the accumulation of the bond stresses is greater. Hence, the prestress σ_p in the tendon and the compression σ_c in the concrete also increase. When x reaches a certain length l_{tr} (the distance between sections a and b in Figure 9-8a, the bond stress within the distance l_{tr} will be in equilibrium with the tension $\sigma_p A_p$ in the tendon. Therefore, the required effective prestress σ_{pe} in the tendon and σ_{pc} in concrete can be established only for x beyond the length l_{tr}, that is, beyond section b in Figure 9-8(c). This length l_{tr} is therefore called the transfer length of the tendon and the portion of the member between sections a and b is called the self-anchoring zone of the pretensioned member. Due to low prestress in the self-anchoring zone, for calculations on shear bearing capacity of oblique section and resistance to cracking of normal section and oblique section at the end of the member, the change of the actual stress value of prestressed tendon in the range of transfer length l_{tr} should be considered. A linear variation of prestress in the self-anchoring zone can be assumed in calculation. As shown in the dotted line in Figure 9-8(c), the end of the component is zero and the end of the prestress transfer length is the value of effective prestress σ_{pe}. The stress distribution along the transfer length is given by the Eq. (9-17).

$$l_{tr} = \alpha \frac{\sigma_{pe}}{f'_{tk}} d \tag{9-17}$$

Where σ_{pe}——effective prestress of tendon;

d——nominal diameter of the tendon, according to Table 2-1 and 2-3 in Appendix 2;

α——tendon's shape coefficient (Table 2-1);

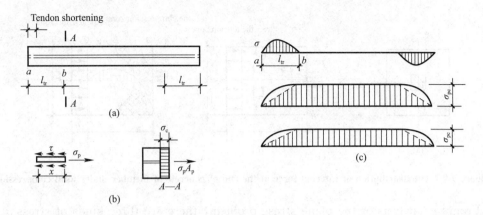

Figure 9-8 The transfer of prestress

(a) The retraction of tendon when relaxing tendon; (b) The distribution of bonding stress τ on the surface of tendon and the stress on the section A—A; (c) The distribution of bonding stress, tensile stress of tendon and pre-compressive stress of concrete along the length of member

f'_{tk}——axial cube tensile strength of concrete in accordance with the standard values of compressive strength at the time of prestress transfer, which can be determined by linear interpolation method in Table 1-2 in Appendix 1.

9.1.9 Calculation for Local Compression Bearing Capacity at the End Anchorage Zone of the Posttensioning Members

The pre-compression of posttensioning member is transferred to concrete with anchorage through the plate. The pre-compression is often large, yet the contact area between plate and concrete under anchorage is often very small, so concrete under anchorage will bear large local pressure. Under the action of local pressure, when the strength or deformation capacity of concrete is insufficient, the component end may produce cracks, even to the local compression damage. The distribution of internal force at the end of concrete is shown in Figure 9-9 when the concrete bears local compression.

According to elastic mechanics, a period of distance is required if local compression under anchorage spreads to the whole cross section. Therefore, the total pre-compression N_P, exerting on the area A_l of cross section AB shown in Figure 9-9(a) and (b), gradually spreads to a larger cross section, with a certain spreading distance until the stress is uniform on the entire section. The section from the end of local compression to the uniform compression distribution of the entire section is known as anchorage zone of prestressed concrete member, which is the section $ABDC$ in Figure 9-9(c). The experimental study shows that the length of the anchorage zone is approximately equal to the height h of the cross section.

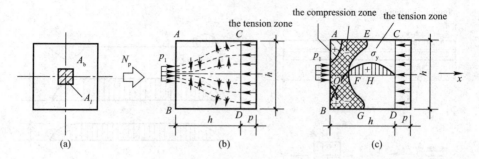

Figure 9-9 The distribution of internal force at the end of concrete of member under local compression

From the analysis of the plane stress problem, there are three kinds of stress σ_x, σ_y and τ at any point in the anchorage zone. σ_x is the normal stress along x direction (longitudinal direction) and most of σ_x in the block $ABDC$ is compressive stress. The value is bigger along the axis O_x, in which the maximum p_1 is at the point O. σ_y is the normal stress along y direction (transverse direction). σ_y in the block $AOBGFE$ is compressive stress and in the block $EFGDC$ is tensile stress. The maximum transverse tensile stress occurs at H (Figure 9-9c). When the pre-compression N_P gradually increases, the tensile strain at H exceeds the ultimate tensile strain of concrete, leading to longitudinal cracks in concrete and lack of bearing capacity as local compression failure. Therefore, *Code for Design of Concrete Structures* regulates that the design should ensure that concrete in anchorage zone is capable to resist cracking or excessive distortion as tensioning the prestressed tendons. The number of indirect reinforcement in anchorage zone should be calculated to meet the requirement of the local bearing capacity.

1. Sectional size of local compression area of member

The test shows that concrete under the locally compressed plate will produce excessive subsidence deformation when the local compression area has too much reinforcement. To limit the deformation, sectional size of local compression zone should meet the following requirements for the concrete structures configuring indirect reinforcement:

$$F_l \leqslant 1.35\beta_c\beta_l f_c A_{ln} \tag{9-18}$$

$$\beta_l = \sqrt{\frac{A_b}{A_l}} \tag{9-19}$$

Where F_l——design value of local load or local compression acting on local compression surface; for anchor head pressure zone of concrete members with bonded prestress, $F_l = 1.2\sigma_{con}A_p$ is taken;

f_c——design value of axial compressive strength of concrete. At the tension stage for post-tensioned prestressed concrete member, according to the cube compressive strength f'_{cu} value of concrete in the corresponding stage, f_c can be calculated by linear interpolation method in Table 1-3 of Appendix 1;

β_c——influence coefficient of concrete strength: when the strength grade of concrete is not more than C50, $\beta_c=1.0$; when the concrete strength grade is equal to C80, $\beta_c=0.8$, between which the linear interpolation method is taken;

β_l——strength enhancement factor of concrete under local compression;

A_{ln}——net area of local compression of concrete: for the post-tensioned members, the area of the holes and grooves should be deducted from the local compression area of the concrete;

A_b——calculated bottom area with local compression, which can be determined by the concentric and symmetrical principle between the local compression area and the calculated bottom area, Figure 9-10 show the common situations;

A_l——local compression area of concrete; the pre-compression is diffused rigidly along the plate by 45°diffusion angle, as shown in Figure 9-11.

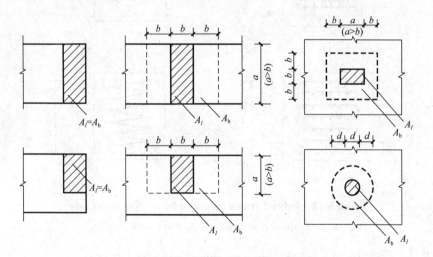

Figure 9-10 The calculated bottom area A_b under local compression

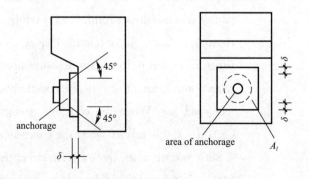

Figure 9-11 The compressive area under prestress diffusion

The section size of the end anchorage zone should be increased to adjust the anchor position or the strength grade of concrete should be improved when Eq. (9-18) is not satisfied.

2. Calculation for bearing capacity under local compression

The indirect reinforcement anchorage zone (welded wire or spiral reinforcement) can prevent local compression failure. When the grid or indirect spiral reinforcement is configured with the core area $A_{cor} \geqslant A_l$ (Figure 9-12), local compression bearing capacity should be calculated by the following equations:

$$F_l \leqslant 0.9(\beta_c\beta_l f_c + 2\alpha\rho_v\beta_{cor} f_{yv})A_{ln} \qquad (9\text{-}20)$$

$$\beta_{cor} = \sqrt{\frac{A_{cor}}{A_l}} \qquad (9\text{-}21)$$

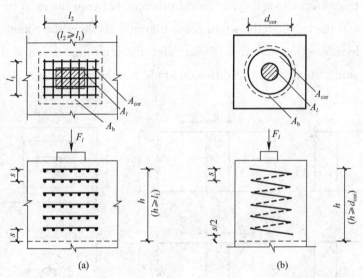

Figure 9-12 Indirect reinforcement in local compression zone

(a) Grid type of reinforcement; (b) Spiral reinforcement

Where F_l, β_c, β_l, f_c, A_{ln}——the same meanings as in Eq. (9-18);

β_{cor}——improvement coefficient of local compression bearing capacity when indirect reinforced is configured, for A_{cor} larger than A_b, $A_{cor} = A_b$ is taken; For A_{cor} not more than 1.25 times of concrete local compression area A_l, $\beta_{cor} = 1.0$;

α——reduction coefficient of concrete constraints resulting from indirect steel bar. When the concrete strength is not more than C50, $\alpha = 1.0$ is taken; when the concrete strength is C80, $\alpha = 0.85$ is taken; when the concrete strength is between C50 and C80, α is determined by linear interpolation;

A_{cor}——concrete core sectional area (duct area is included) within the inner surface when grid or spiral indirect steel bars are configured, which should be greater than the concrete local compression area A_l. Its center of gravity should be coincident with A_l, calculated according to principles of concentric symmetry;

ρ_v——volume reinforcement ratio of indirect rebar (volume of indirect reinforcement contained per unit volume within the core area A_{cor}), $\rho_v \geqslant 0.5\%$ is required.

For grid reinforcement (Figure 9-12a),

$$\rho_v = \frac{n_1 A_{s1} l_1 + n_2 A_{s2} l_2}{A_{cor} s} \quad (9-22)$$

At this point, the ratio between the section areas of steel within unit length in both directions of the steel mesh should not be greater than 1.5.

For spiral reinforcement (Figure 9-12b),

$$\rho_v = \frac{4 A_{ss1}}{d_{cor} s} \quad (9-23)$$

Where n_1, A_{s1}——number of bars along the l_1 direction for grid and the sectional area of the single bar;

n_2, A_{s2}——number of bars along the l_2 direction for grid and the sectional area of the single bar;

A_{ss1}——the sectional area of one spiral indirect steel;

d_{cor}——concrete cross section diameter within the inner surface of spiral direct reinforcement;

s——the spacing of the grid or spiral indirect reinforcement with the common value of 30~80mm.

Indirect reinforcement calculated by Eq. (9-20) should be configured within height h (Figure 9-12). It should be not less than 4 sheets for the grid reinforcement, and not less than 4 circles for spiral steel.

If Eq. (9-20) is not satisfied, the increase on the number of bars, the diameter and reduction on the spacing of steel nets should be conducted for grid reinforcement nets; for spiral reinforcement, the diameter should be increased and the pitch distance should be decreased.

9.2 Axial Tension Members of Prestressed Concrete

9.2.1 Stress Analysis of Members Subjected to Axial Tension

For prestressed concrete axial tensile member from tensioning prestressed tendon to failure, it can be divided into two phases for the change of stress in concrete and prestressed tendons: construction phase and service phase. Each stage includes a number of features during loading process. Therefore, when designing prestressed concrete members, mechanical performances including bearing capacity, crack resistance should be checked under all phases.

1. Pretensioned members

Figure 9-13 is the diagram at each stage for pretensioned and posttensioned prestressed concrete structure under axial tension.

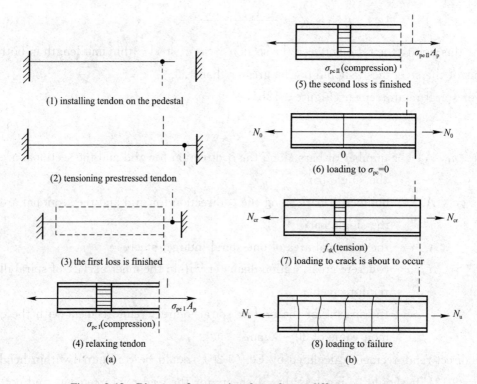

Figure 9-13 Diagram of pretensioned member at different stress stages
(a) Construction stage; (b) Service stage

(1) Construction phase

1) Tensioning prestressed tendons. The prestressed tendons of section area A_p is tensioned to its control stress σ_{con} at the pedestal and the total tension for prestressed tendon is $\sigma_{con}A_p$ at the point. Ordinary steel does not support any stress.

2) Before the concrete under pre-compressive stress, the first loss is completed. After the tensioning prestressed tendons is completed, prestressed tendons are anchored in the pedestal, followed by pouring concrete and steam-curing. The first batch of prestress loss σ_{lI} is generated by anchorage deformation, temperature difference and relaxation part of

prestressed tendons. Tensile stress σ_{con} of prestressed tendon is reduced to $\sigma_{pe} = \sigma_{con} - \sigma_{l\mathrm{I}}$. At this point, since the tendons have not relaxed, the concrete stress is $\sigma_{pc} = 0$ and ordinary steel stress is $\sigma_s = 0$.

3) Releasing prestressed tendons. When the concrete reaches over 75% of the strength design values, the prestressed tendon is released with retraction. Concrete is under compression dependent on the bond strength between concrete and prestressed tendon. Since prestressed tendon is shortened, the tension wil reduce. Precompressive stress of concrete is defined as $\sigma_{pc\mathrm{I}}$ obtained by relaxation of prestressed tendon. Due to the deformation of both tendons and concrete, tensile stress of prestressed tendon decreases $\alpha_E \sigma_{pc\mathrm{I}}$. That is:

$$\sigma_{pe\mathrm{I}} = \sigma_{con} - \sigma_{l\mathrm{I}} - \alpha_E \sigma_{pc\mathrm{I}} \qquad (9\text{-}24)$$

Meanwhile, the steel bar has also obtained precompressive stress $\sigma_{s\mathrm{I}}$

$$\sigma_{s\mathrm{I}} = \alpha_E \sigma_{pc\mathrm{I}}$$

Where α_E——the ratio of the elastic modulus between prestressed tendons or steel bar and concrete, $\alpha_E = \dfrac{E_s}{E_c}$.

Obtained by the force equilibrium conditions

$$\sigma_{pe\mathrm{I}} A_p = \sigma_{pc\mathrm{I}} A_c + \sigma_{s\mathrm{I}} A_s$$

Then

$$\sigma_{pc\mathrm{I}} = \frac{(\sigma_{con} - \sigma_{l\mathrm{I}})A_p}{A_c + \alpha_E A_s + \alpha_E A_p} = \frac{N_{p\mathrm{I}}}{A_n + \alpha_E A_p} = \frac{N_{p\mathrm{I}}}{A_0} \qquad (9\text{-}25)$$

Where A_c——concrete section area deducting the area of prestressed tendons and the steel bar;

A_0——conversion section area of component;

A_n——net section area of component;

$N_{p\mathrm{I}}$——total pretension of prestressed tendons after completing the first loss, $N_{p\mathrm{I}} = (\sigma_{con} - \sigma_{l\mathrm{I}}) A_p$.

4) Concrete produces pre-compressive stress after the second loss is finished. With time growing, due to further relaxation for prestressed tendon, concrete occurs shrinkage, creep, which leads to a second batch of prestressed loss as $\sigma_{l\mathrm{II}}$. At this point, concrete and rebar will be further shortened, compressive stress of concrete reduces from $\sigma_{pc\mathrm{I}}$ to $\sigma_{pc\mathrm{II}}$, tensile stress of prestressed reinforcement reduces from $\sigma_{pe\mathrm{I}}$ to $\sigma_{pe\mathrm{II}}$ and compressive stress of steel bars reduces to $\sigma_{s\mathrm{II}}$, so

$$\begin{aligned}\sigma_{pe\mathrm{II}} &= \sigma_{con} - \sigma_{l\mathrm{I}} - \alpha_E \sigma_{pc\mathrm{I}} - \sigma_{l\mathrm{II}} + \alpha_E(\sigma_{pc\mathrm{I}} - \sigma_{pc\mathrm{II}}) \\ &= \sigma_{con} - \sigma_l - \alpha_E \sigma_{pc\mathrm{II}}\end{aligned} \qquad (9\text{-}26)$$

Where $\alpha_E(\sigma_{pc\mathrm{I}} - \sigma_{pc\mathrm{II}})$——the added value for tensile stress in prestressed tendon caused by the difference value due to reduction on the concrete compressive stress and the recovery of elastic compression of components.

The equation can be obtained by the force equilibrium conditions:
$$\sigma_{peII} A_p = \sigma_{pcII} A_c + \sigma_{sII} A_s$$

At this point, compressive stress σ_{sII} caused by steel bar only includes $\alpha_E \sigma_{pcII}$, which should consider the prestress loss σ_{l5} due to shrinkage and creep of concrete, thus:
$$\sigma_{sII} = \alpha_E \sigma_{pcII} + \sigma_{l5} \tag{9-27}$$

Then
$$\sigma_{pcII} = \frac{(\sigma_{con} - \sigma_l) A_p - \sigma_{l5} A_s}{A_c + \alpha_E A_s + \alpha_E A_p} = \frac{N_{pII} - \sigma_{l5} A_s}{A_0} \tag{9-28}$$

Where σ_{pcII} ——effective preloading stress established in prestressed concrete;

σ_{l5} ——stress of common reinforced steel caused by shrinkage and creep of concrete;

N_{pII} ——total pretension of prestressed tendon after completing all losses, $N_{pII} = (\sigma_{con} - \sigma_l) A_p$.

(2) Loading phase

1) Loading until stress of concrete is zero. Tension of concrete produced by axial force N_0 will offset effective precompressive stress σ_{pcII} of concrete, of which is the decompression, namely $\sigma_{pc} = 0$. At this time, the tensile stress σ_{p0} of prestressed tendon increases $\alpha_e \sigma_{pcII}$ on the basis of σ_{peII}, that is,
$$\sigma_{p0} = \sigma_{peII} + \alpha_E \sigma_{pcII}$$

Taking Eq. (9-26) into the formula, then:
$$\sigma_{p0} = \sigma_{con} - \sigma_l \tag{9-29}$$

Compressive stress σ_s of ordinary steel is added by a tensile stress $\alpha_E \sigma_{pcII}$ on the basis of previous compressive stress σ_{sII}, so
$$\sigma_s = \sigma_{sII} - \alpha_E \sigma_{pcII} = \alpha_E \sigma_{pcII} + \sigma_{l5} - \alpha_E \sigma_{pcII} = \sigma_{l5}$$

The stress for ordinary reinforcement is still under compression at this stage from above equations, and the value of which is σ_{l5}.

Axial tensile force N_0 can be obtained by the force equilibrium conditions:
$$N_0 = \sigma_{p0} A_p - \sigma_{l5} A_s = (\sigma_{con} - \sigma_l) A_p - \sigma_{l5} A_s$$
$$= N_{pII} - \sigma_{l5} A_s$$

By Eq. (9-28):
$$N_{pII} - \sigma_{l5} A_s = \sigma_{pcII} A_0$$

Therefore
$$N_0 = \sigma_{pcII} A_0 \tag{9-30}$$

Where N_0 ——axial tensile force as stress of concrete is zero.

2) Loading until the cracks appear. After the axial force exceeds N_0, concrete is in

tension. As the load increases, the tensile stress is also growing. When the load is increased to N_{cr} that the tensile stresses of concrete reach the standard value of axis tensile strength f_{tk}, concrete will appear cracks. At this point, tensile stress of prestressed tendon σ_{pcr} should add $\alpha_E f_{tk}$ based on σ_{p0}, that is:

$$\sigma_{pcr} = \sigma_{p0} + \alpha_E f_{tk} = \sigma_{con} - \sigma_l + \alpha_E f_{tk}$$

Stress of steel bar σ_s is transferred into tensile stress from compressive stress, and the value is

$$\sigma_s = \alpha_E f_{tk} - \sigma_{l5}$$

Axial tensile force N_{cr} can be obtained by the force equilibrium conditions:

$$N_{cr} = \sigma_{pcr} A_p + \sigma_s A_s + f_{tk} A_0$$

Taking σ_{pcr}, σ_s into the above equation:

$$N_{cr} = (\sigma_{pc\mathrm{II}} + f_{tk}) A_0 \tag{9-31}$$

Obviously, due to the pre-compressive stress $\sigma_{pc\mathrm{II}}$ ($\sigma_{pc\mathrm{II}}$ is much larger than f_{tk}), N_{cr} for prestressed concrete is much higher than axial tension of ordinary reinforced concrete. This is why the crack resistance of prestressed concrete members is high.

3) Load to failure. When the axial force exceeds N_{cr}, concrete works with cracks. Tension is all supported by prestressed tendons and steel bars. Stress of prestressed tendons and steel bars will reach tensile strength design values of f_{py} and f_y at failure.

Axial tensile force N_u can be obtained by the force equilibrium conditions:

$$N_u = f_{py} A_p + f_y A_s \tag{9-32}$$

2. Posttensioned members

Figure 9-14 is the diagram at each stage for posttensioned prestressed concrete structures under axial tension.

(1) Construction phase

1) After pouring concrete, the stage before the prestressed tendon is tensioned can be considered that there is no stress in the cross section;

2) Tensioning prestressed tendons. At the same time of tensioning prestressed tendons, jack's reaction transfers to concrete through the force transmission frame, leading to elastic compression in concrete and friction losses σ_{l2}. At this point, tensile stress of prestressed tendon is $\sigma_{pe} = \sigma_{con} - \sigma_{l2}$. Compressive stress of steel bar is $\sigma_s = \alpha_E \sigma_{pc}$.

Pre-compressive stress of concrete σ_{pc} can be obtained by the force equilibrium conditions:

$$\sigma_{pe} A_p = \sigma_{pc} A_c + \sigma_s A_s$$

σ_{pe} and σ_s are substituted in the above equation,

$$(\sigma_{con} - \sigma_{l2}) A_p = \sigma_{pc} A_c + \alpha_E \sigma_{pc} A_s$$

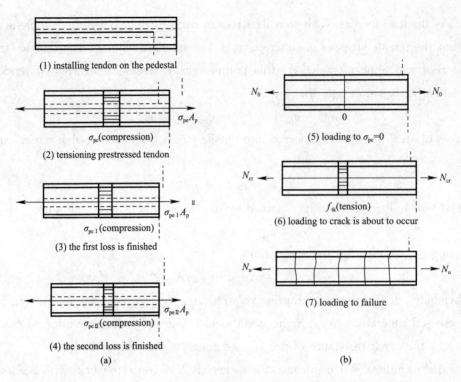

Figure 9-14 Diagram of post-tensioned member at different stress stages
(a) Construction stage; (b) Service stage

$$\sigma_{pc} = \frac{(\sigma_{con} - \sigma_{l2})A_p}{A_c + \alpha_E A_s} = \frac{(\sigma_{con} - \sigma_{l2})A_p}{A_n}$$

Where A_c——section area of concrete deducting the section area of steel bar and the reserved channel.

3) Before the concrete is under pre-compressive stress, the first loss is completed. After tensioning prestressed tendon, stress loss σ_{l1} is caused by anchorage deformation and retraction of tendon. At this time, tensile stress of prestressed tendon reduces to $\sigma_{con} - \sigma_{l2} - \sigma_{l1}$ from $\sigma_{con} - \sigma_{l2}$, so

$$\sigma_{pe\,I} = \sigma_{con} - \sigma_{l2} - \sigma_{l1} = \sigma_{con} - \sigma_{l\,I} \tag{9-33}$$

Compressive stress of steel bar is $\sigma_{s\,I} = \alpha_E \sigma_{pc\,I}$.

Compressive stress of concrete $\sigma_{pc\,I}$ can be obtained by the force equilibrium conditions:

$$\sigma_{pe\,I} A_p = \sigma_{pc\,I} A_c + \sigma_{s\,I} A_s$$

$\sigma_{pe\,I}$, $\sigma_{s\,I}$ are substituted in the above equation,

$$(\sigma_{con} - \sigma_{l\,I})A_p = \sigma_{pc\,I} A_c + \alpha_E \sigma_{pc\,I} A_s$$

$$\sigma_{pc\,I} = \frac{(\sigma_{con} - \sigma_{l1})A_p}{A_c + \alpha_E A_s} = \frac{N_{p\,I}}{A_n} \tag{9-34}$$

4) After concrete is under the pre-compressive stress, the second batch of loss is completed. Because of the relaxation, shrinkage and creep of prestressed tendon cause loss of stress σ_{l4}, σ_{l5}

(and σ_{l6}), tensile stress of prestressed tendon reduces to σ_{peII} from σ_{peI}, that is

$$\sigma_{peII} = \sigma_{con} - \sigma_{lI} - \sigma_{lII} = \sigma_{con} - \sigma_{l}$$

Compressive stress in the steel bar is

$$\sigma_{sII} = \alpha_E \sigma_{pcII} + \sigma_{l5}$$

Compressive stress of concrete σ_{pcII} can be obtained by the force equilibrium conditions:

$$\sigma_{peII} A_p = \sigma_{pcII} A_c + \sigma_{sII} A_s$$

σ_{peII}, σ_{sII} in the above equation are substituted,

$$(\sigma_{con} - \sigma_l) A_p = \sigma_{pcII} A_c + (\alpha_E \sigma_{pcII} + \sigma_{l5}) A_s$$

$$\sigma_{pcII} = \frac{(\sigma_{con} - \sigma_l) A_p - \sigma_{l5} A_p}{A_c + \alpha_E A_s} = \frac{(\sigma_{con} - \sigma_l) A_p - \sigma_{l5} A_p}{A_n} \quad (9\text{-}35)$$

$$= \frac{N_{pII} - \sigma_{l5} A_s}{A_n}$$

(2) Loading stage

1) Loading until stress of concrete is zero. Tensile stress of concrete caused by axial force N_0 will all offset effective pre-compressive stress of concrete σ_{pcII}, of which is the decompression force, namely $\sigma_{pc} = 0$. At this time, tensile stress of prestressed tendon σ_{p0} increases $\alpha_e \sigma_{pcII}$ on the basis of σ_{peII}. That is

$$\sigma_{p0} = \sigma_{peII} + \alpha_E \sigma_{pcII} = \sigma_{con} - \sigma_l + \alpha_E \sigma_{pcII}$$

Stress of steel bar σ_s is added by a tensile stress $\alpha_E \sigma_{pcII}$ on the basis of previous compressive stress $\alpha_E \sigma_{pcII} + \sigma_{l5}$, so

$$\sigma_s = \sigma_{sII} - \alpha_E \sigma_{pcII} = \alpha_E \sigma_{pcII} + \sigma_{l5} - \alpha_E \sigma_{pcII} = \sigma_{l5}$$

The axial tensile force N_0 can be obtained by the force equilibrium conditions:

$$N_0 = \sigma_{p0} A_p - \sigma_{l5} A_s = (\sigma_{con} - \sigma_l + \alpha_E \sigma_{pcII}) A_p - \sigma_{l5} A_s \quad (9\text{-}36a)$$

By Eq. (9-35):

$$(\sigma_{con} - \sigma_l) A_p - \sigma_{l5} A_s = \sigma_{pcII} (A_c + \alpha_E A_s)$$

Therefore,

$$N_0 = \sigma_{pcII} (A_c + \alpha_E A_s) + \alpha_E \sigma_{pcII} A_p$$
$$= \sigma_{pcII} (A_c + \alpha_E A_s + \alpha_E A_p) = \sigma_{pcII} A_0 \quad (9\text{-}36b)$$

2) Loading until the cracks appear. At this time, concrete is in tension, the tensile stress reaches f_{tk} and tensile stress of prestressed tendon σ_{pcr} increases $\alpha_E f_{tk}$ on the basis of σ_{p0}, that is:

$$\sigma_{pcr} = \sigma_{p0} + \alpha_E f_{tk} = (\sigma_{con} - \sigma_l + \alpha_E \sigma_{pcII}) + \alpha_E f_{tk}$$

Stress of steel bar σ_s is transferred into tension from compressive stress σ_{l5}, of which the value is $\sigma_s = \alpha_E f_{tk} - \sigma_{l5}$.

Axial tensile force N_{cr} can be obtained by the force equilibrium conditions:

$$N_{cr} = \sigma_{pcr} A_p + \sigma_s A_s + f_{tk} A_c$$

σ_{pcr}, σ_s in the above equation are substituted

$$N_{cr} = (\sigma_{con} - \sigma_l + \alpha_E \sigma_{pcII} + \alpha_E f_{tk})A_p + (\alpha_E f_{tk} - \sigma_{l5})A_s + f_{tk}A_c$$
$$= (\sigma_{con} - \sigma_l + \alpha_E \sigma_{pcII})A_p - \sigma_{l5}A_s + f_{tk}(A_c + \alpha_E A_s + \alpha_E A_p)$$

From the equal relationship between (9-36a) and (9-36b), that is,

$$N_0 = \sigma_{pcII} A_0 = (\sigma_{con} - \sigma_l + \alpha_E \sigma_{pcII})A_p - \sigma_{l5}A_s$$

Thus

$$N_{cr} = \sigma_{pcII} A_0 + f_{tk} A_0 = (\sigma_{pcII} + f_{tk})A_0 \tag{9-37}$$

3) Loading to failure. Similar with pre-tensioned members, tensile stress of tendons and steel bars respectively reach f_{py} and f_y. N_u can be obtained from force equilibrium conditions:

$$N_u = f_{py}A_p + f_y A_s \tag{9-38}$$

From the above analysis, we have the following conclusions:

(1) During the construction phase, the equations of σ_{pcII} are basically the same between pretensioned and posttensioned members, but σ_l are different. In addition, the conversion section for pretensioned members is A_0 while for posttensioned member is A_n. Effective compression stress of posttensioned components σ_{pcII} is higher.

(2) At service phase, the forms of three formulas N_0, N_{cr}, N_u are the same regardless of pre-tensioned or posttensioned members, but σ_{pcII} are not the same when calculating N_0 and N_{cr}.

(3) The tendon stress is always at high levels. In addition, the variation in tendon stress before cracking of concrete is very small. Moreover, the concrete is subjected to compression before the load reaches N_0. Therefore, the advantages of both materials are fully utilized.

(4) The cracking for a prestressed member is usually much later than the cracking of a non-prestressed member, with the accompanying phenomenon that the cracking load is closer to the ultimate load. The latter phenomenon may show negative effects in certain cases.

(5) The ultimate bearing capacity of a prestressed member is equal to the ultimate bearing capacity of the corresponding non-prestressed member with the same materials and section dimensions.

9.2.2 Calculation for Axial Tension Components at Service Stage

In the design of a prestressed member, the whole process from construction to serviceand finally to ultimate state should be considered. Therefore, the design can also be divided into construction stage and loading stage. These include the calculation of ultimate bearing capacity as well as separate checking in different stages on crack resistance, crack width, bearing capacity and local bearing strength of the anchorage zone (if applicable) during tensioning or releasing tendons.

1. Calculation of bearing capacity for the loading stage

Section calculation diagram is shown in Figure 9-15(a), the axial tensile load in the normal section of member is calculated according to the following equation:

$$N \leqslant N_u = f_{py}A_p + f_y A_s \tag{9-39}$$

Where N——design value of axial tension;

f_{py}, f_y——design values of tensile strength for prestressed tendon and steel bar;

A_p, A_s——the whole section areas of longitudinal prestressed tendon and steel bar.

2. Checking on crack resistance and crack width

According to Eqs. (9-31) and (9-37), if the axial force N is not more than N_{cr}, the component will be not cracked. Its calculation diagram can be shown in Figure 9-15(b).

$$N \leqslant N_{cr} = (\sigma_{pcII} + f_{tk})A_0 \tag{9-40}$$

The above equation can be expressed in the form of stress, namely

$$\frac{N}{A_0} \leqslant \sigma_{pcII} + f_{tk} \tag{9-41}$$

$$\sigma_c - \sigma_{pcII} \leqslant f_{tk}$$

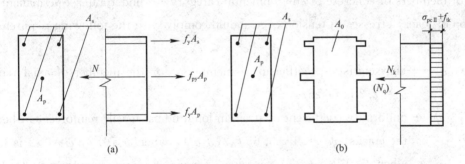

Figure 9-15 Calculated scheme of bearing capacity for prestressed member of axial tension at the service stage

(a) Calculated scheme of bearing capacity; (b) Checking scheme for cracks

For different prestressed components, there are different crack security reserves according to requirements of environment and service. Three levels for crack control of normal section for prestressed concrete members are shown as follows:

(1) The first level——members are strictly not to be cracked

Component edges in tension should not appear tensile stress under standard combinations of loads:

$$\sigma_{ck} - \sigma_{pcII} \leqslant 0 \tag{9-42}$$

(2) The second level——members are not commonly allowed to be cracked

The stress of component edges in tension should not be more than standard value of tensile strength of concrete under standard combinations of loads:

$$\sigma_{ck} - \sigma_{pc\,II} \leqslant f_{tk} \qquad (9\text{-}43)$$

Where σ_{ck}——normal stress of concrete in crack resistance checking under the standard combinations of loads;

$$\sigma_{ck} = \frac{N_k}{A_0} \qquad (9\text{-}44)$$

Where N_k——axial load value under the standard combinations of loads;

A_0——area of transformed section, $A_0 = A_c + \alpha_E A_p + \alpha_E A_s$;

$\sigma_{pc\,II}$——the pre-compressive stress of concrete in edges of crack resistance checking after deducting the whole prestress losses, which is calculated according to Eq. (9-28);

f_{tk}——characteristic value of axial tensile strength of concrete.

(3) The third level——members are allowed to be cracked

The maximum crack width is calculated under the standard combinations of loads with consideration on the long-term effects, which should meet the following requirements:

$$w_{max} = \alpha_{cr} \psi \frac{\sigma_s}{E_s} \left(1.9 c_s + 0.08 \frac{d_{eq}}{\rho_{te}}\right) \leqslant w_{lim} \qquad (9\text{-}45)$$

For members of concrete of environmental category 2-a under a quasi-permanent combination of loads, stresses in tension edge should comply with the following requirements:

$$\sigma_{cq} - \sigma_{pc\,II} \leqslant f_{tk} \qquad (9\text{-}46)$$

Where α_{cr}——characteristics coefficient of members, for the member of axial tension, $\alpha_{cr} = 2.2$ is taken;

ψ——uniformity coefficient of strain in longitudinal tensile reinforcement between the cracks, $\psi = 1.1 - 0.65 f_{tk}/(\rho_{te}\sigma_s)$, when $\psi < 0.2$, $\psi = 0.2$ is taken; when $\psi > 1.0$, $\psi = 1.0$ is taken; for direct bearing component, $\psi = 1.0$ is taken;

ρ_{te}——reinforcement ratio of longitudinal tensile tendon according to calculation of effective tensile sectional area of concrete, $\rho_{te} = (A_s + A_p)/A_{te}$, when $\rho_{te} < 0.01$, $\rho_{te} = 0.01$ is taken;

A_{te}——effective tensile section area of concrete, $A_{te} = bh$;

σ_s——equivalent stress of longitudinal tensile reinforcement in prestressed concrete members under standard combination of loads;

$$\sigma_s = \frac{N_k - N_{p0}}{A_p + A_s}$$

σ_{cq}——normal stress at the concrete edge for crack resistance checking under quasi- permanent combination;

N_{p0}——the axial force when normal prestress of concrete is zero on the calculation section;

c_s——the distance between the outer edge of longitudinal tensile reinforcement and the bottom edge of tensile zone(mm). When $c<20$, $c=20$ is taken; when $c>65$, $c=65$ is taken;

A_p, A_s——section areas of longitudinal prestressed tendons and common steel bars in tension zone;

d_{eq}——equivalent diameter of longitudinal reinforcement in tension zone (mm);

$$d_{eq} = \frac{\sum n_i d_i^2}{\sum n_i v_i d_i} \tag{9-47}$$

d_i——nominal diameters of the ith longitudinal reinforcement in tension zone (mm);

n_i——number of the ith longitudinal reinforcements in tension zone;

v_i——relative bonding characteristic coefficient of the ith longitudinal reinforcement in tension zone, which can be taken by Table 9-5;

ω_{lim}——limit value of the maximum crack width.

Relative bonding characteristic coefficient for steel bars   Table 9-5

Reinforced categories	Steel bar		Pretensioned prestressed tendons			Posttensioned prestressed tendons		
	Plain round bar	Ribbed	Ribbed	Spiral rib steel wire	Steel cable	Ribbed	Steel cable	Round wire
v_i	0.7	1.0	1.0	0.8	0.6	0.8	0.5	0.4

Note: For epoxy-coated rebar, the value is 0.8 times of the value in the above table.

9.2.3 Checking for Axial Tensile Members at Construction Stage

Concrete will reach the maximum compressive stress σ_{cc} after pretensioning or posttensioning, yet stress of concrete reaches only 75% of design strength at this time. It should check strength of members, which include two parts as follows.

1. Checking on bearing capacity of components when tensioning (or relaxing) prestressed tendons

In order to ensure concrete won't be crushed when tensioning (or relaxing) prestressed tendon, the compressive stress of concrete should meet the following conditions:

$$\sigma_{cc} \leqslant 0.8 f'_{ck} \tag{9-48}$$

Where f'_{ck}——standard values of axial compressive strength, which is relative to cube compressive strength of concrete f'_{cu} when tensioning (or relaxing) prestressed tendon. This value can be taken by linear interpolation according to Table 1-1 in Appendix 1.

In the case of pre-tensioned members, the first batch of prestress loss is included in the calculation of σ_{cc}. Therefore,

$$\sigma_{cc} = \frac{(\sigma_{con} - \sigma_{l1})A_p}{A_0} \tag{9-49}$$

In the case of posttensioned members, no prestress loss is considered in the calculation of σ_{cc}. Therefore,

$$\sigma_{cc} = \frac{\sigma_{con} A_p}{A_n} \tag{9-50}$$

2. Checking on local bearing strength of the anchorage zone

The checking is based on Eqs. (9-18) and (9-20).

3. The whole procedure for design prestressed member subjected to axial tension

The steps for the design of a prestressed member subjected to axial tension are as follows:

(1) The follow variables should be determined: the section dimension, the strengths and the elastic modulus of concrete, tendons and rebars, the percentage of concrete strength at prestress transfer, the tendon control stress, the construction method (pretensioning or posttensioning), the internal forces caused by external loading, the coefficient of structural importance, etc;

(2) A_p and A_s are to be determined according to the ultimate bearing capacity requirement;

(3) The prestress loss σ_l is calculated;

(4) The effective compressive prestress σ_{pcII} is calculated;

(5) The crack resistance or the crack width under service load conditions is to be checked, if it is not satisfied, the adjustment on initial parameters is undertaken to go back to step 1;

(6) Checking on the construction stage. For pretensioned members, the concrete strength requirement at tendon releasing is to be checked; for posttensioned members, the concrete strength requirement and the local bearing strength requirement at jacking are to be checked. If any of these are not satisfied, the adjustment on the initial parameters should be undertaken on step 1 until all the requirements are satisfied.

[Example 9-1] The calculation of bottom chord for a 24m prestressed concrete truss is required. The information and data is shown in Table 9-6.

Design conditions   Table 9-6

Materials	Concrete	Tendon	Steel
Category and strength grade	C60	Steel strand	HRB400
Section	280mm×180mm holes 2ϕ55	4ϕ^s1×7 (d=15.2mm)	According to the configuration requirements 4Φ12 (A_s=452mm^2)
Material strength (N/mm^2)	f_c=27.5, f_{ck}=38.5 f_t=2.04, f_{tk}=2.85	f_{ptk}=1860 f_{py}=1320	f_{yk}=400 f_y=360
Elastic modulus (N/mm^2)	E_c=3.6×10^4	E_s=1.95×10^5	E_s=2×10^5
Tension control stress	σ_{con}=0.70f_{ptk}=0.70×1860=1302N/mm^2		
Concrete strength under tensioning	f'_{cu}=60N/mm^2		
Tensioning technology	Posttensioning, tension at one end, adopting wedge anchorage, and the hole is an embedded plastic corrugated pipe		
Internal force of member	Axial tension caused by permanent load standard value is N_k=820kN, Axial tension caused by variable load standard value is N_k=320kN, Quasi-permanent value coefficient of variable load is 0.5		
Structural importance coefficient	γ_0=1.1		

[Solution]

(1) Calculation of bearing capacity at service stage

From Eq. (9-39)

$$A_p = \frac{\gamma_0 N - f_y A_s}{f_{py}}$$

$$= \frac{1.1 \times (1.2 \times 820 \times 10^3 + 1.4 \times 320 \times 10^3) - 360 \times 452}{1320} = 1070 \text{mm}^2$$

Two high-strength low relaxation steel strands are adopted, each of which is equipped with 4ϕ^s1×7 with d=15.2mm (A_p=1120mm^2), as shown in Figure 9-16(c).

(2) Crack resistance checking at service stage

1) Sectional geometry

Prestress $\alpha_{E1} = \dfrac{E_s}{E_c} = \dfrac{1.95 \times 10^5}{3.6 \times 10^4} = 5.42$

Non-prestress $\alpha_{E2} = \dfrac{2.0 \times 10^5}{3.6 \times 10^4} = 5.56$

$$A_n = A_c + \alpha_{E2} A_s = 280 \times 180 - 2 \times \frac{\pi}{4} \times 55^2 - 452 + 5.56 \times 452$$

$$= 47709 \text{mm}^2$$

$$A_0 = A_n + \alpha_{E1} A_p = 47709 + 5.42 \times 1120 = 53779 \text{mm}^2$$

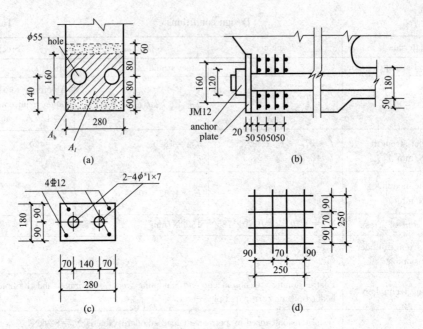

Figure 9-16 Bottom chord of truss(Unit: mm)
(a) Compression area; (b) End node of the bottom chord; (c) Section reinforcement of the bottom chord; (d) Steel mesh

2) Calculation on prestress loss

① Information of loss of anchorage σ_{l1}

According to Table 9-2, for wedge anchorage $a=5$mm, thus

$$\sigma_{l1} = \frac{a}{l}E_s = \frac{5}{24000} \times 1.95 \times 10^5 = 40.63 \text{N/mm}^2$$

② Friction loss σ_{l2}

The loss is calculated according to the anchor end, so $l=24$m, due to linear reinforcement, $\theta=0°$, $kx=0.0015 \times 24=0.036<0.3$. σ_{l2} can be calculated by the following:

$$\sigma_{l2} = (kx+\mu\theta)\sigma_{con} = (0.0015 \times 24) \times 1302 = 46.87 \text{N/mm}^2$$

Therefore, the first batch of loss is

$$\sigma_{l\text{I}} = \sigma_{l1} + \sigma_{l2} = 40.63 + 46.87 = 87.50 \text{N/mm}^2$$

③ Stress relaxation loss of prestressed reinforcement σ_{l4}

$$\frac{\sigma_{con}}{f_{ptk}} = \frac{1302}{1860} = 0.7 > 0.5$$

$$\sigma_{l4} = 0.125\left(\frac{\sigma_{con}}{f_{ptk}} - 0.5\right)\sigma_{con}$$

$$= 0.125\left(\frac{1302}{1860} - 0.5\right) \times 1302 = 32.55 \text{N/mm}^2$$

④ Shrinkage and creep damage of concrete σ_{l5}

$$\sigma_{pc\text{I}} = \frac{(\sigma_{con} - \sigma_{l\text{I}})A_p}{A_n} = \frac{(1302 - 87.50) \times 1120}{47709} = 28.51 \text{N/mm}^2$$

$$\frac{\sigma_{pcI}}{f'_{cu}} = \frac{28.51}{60} = 0.48 < 0.50$$

$$\rho = \frac{A_p + A_s}{A_n} = \frac{1120 + 452}{2 \times 47709} = 0.0165$$

$$\rho_{l5} = \frac{55 + 300 \dfrac{\sigma_{pcI}}{f'_{cu}}}{1 + 15\rho} = \frac{55 + 300 \times 0.48}{1 + 15 \times 0.0165} = 159.52 \text{N/mm}^2$$

Thus, the second batch of loss is

$$\sigma_{lII} = \sigma_{l4} + \sigma_{l5} = 32.55 + 159.52 = 192.07 \text{N/mm}^2$$

The total loss

$$\sigma_l = \sigma_{lI} + \sigma_{lII} = 87.50 + 192.07 = 279.57 \text{N/mm}^2$$

3) Checking on crack resistance

The effective prestress of concrete is to be calculated.

Under the environment of Class 1, the prestressed concrete roof truss is checked according to the second level of crack control degree.

$$\sigma_{pcII} = \frac{(\sigma_{con} - \sigma_l)A_p - \sigma_{l5}A_s}{A_n}$$

$$= \frac{(1302 - 279.57) \times 1120 - 159.52 \times 452}{47709} = 22.49 \text{N/mm}^2$$

Under the standard combination of load

$$N_k = 820 + 320 = 1140 \text{kN}$$

$$\sigma_{ck} = \frac{N_k}{A_0} = \frac{1140 \times 10^3}{53779} = 21.20 \text{N/mm}^2$$

$$\sigma_{ck} - \sigma_{pcII} = 21.20 - 22.49 < 0$$

The requirements are met.

(3) Checking at construction stage

The maximum tension

$$N_p = \sigma_{con} \times A_p = 1302 \times 1120 = 1458000 \text{N} = 1458 \text{kN}$$

Compression of concrete on the section

$$\sigma_{cc} = \frac{N_p}{A_n} = \frac{1458 \times 10^3}{47709} = 30.56 \text{N/mm}^2 < 0.8 f'_{ck}$$

$$= 0.8 \times 38.5 = 30.8 \text{N/mm}^2$$

The requirements are met.

(4) Checking on local compression zone under anchorage

1) Sectional dimension checking at the end of compression zone

The diameter of clamp anchorage is 120mm, and the thickness of plate under anchorage is 20mm. The local compression area can be calculated according to 45°diffusion area of the compression F_l from edge of anchorage in the plate. In the calculation of local compression bottom area, the shaded rectangular area shown in Figure 9-16(a) can approximately replace two-circular area.

$$A_l = 280 \times (120 + 2 \times 20) = 44800 \text{mm}^2$$

Local compression calculating bottom area under anchorage

$$A_b = 280 \times (160 + 2 \times 60) = 78400 \text{mm}^2$$

Local compression net area of concrete

$$A_{ln} = 44800 - 2 \times \frac{\pi}{4} \times 55^2 = 40048 \text{mm}^2$$

$$\beta_l = \sqrt{\frac{A_b}{A_l}} = \sqrt{\frac{78400}{44800}} = 1.323$$

When $f_{cu,k} = 60\text{N/mm}^2$, $\beta_c = 0.933$, according to the linear interpolation method from Eq. (9-18)

$$F_l = 1.2\sigma_{con} A_p = 1.2 \times 1302 \times 1120 = 1749888\text{N} \approx 1749.9\text{kN}$$
$$< 1.35\beta_c\beta_l f_c A_{ln} = 1.35 \times 0.933 \times 1.323 \times 27.5 \times 40048$$
$$= 1835 \times 10^3 \text{N} = 1835\text{kN}$$

The requirements are met.

2) Calculation of local compression bearing capacity

Indirect steel bars are made of 4 pieces of $\phi 8$ grid welded mesh, as shown in Figure 9-16(b). The spacing is $s = 50\text{mm}$ and the mesh size is shown in Figure 9-16(d).

$$A_{cor} = 250 \times 250 = 62500 \text{mm}^2 > A_l = 44800 \text{mm}^2$$

$$\beta_{cor} = \sqrt{\frac{A_{cor}}{A_l}} = \sqrt{\frac{62500}{44800}} = 1.181$$

Volume reinforcement ratio of indirect reinforcement

$$\rho_v = \frac{n_1 A_{s1} l_1 + n_2 A_{s2} l_2}{A_{cor} s} = \frac{4 \times 50.3 \times 250 + 4 \times 50.3 \times 250}{62500 \times 50} = 0.032$$

According to the Eq. (9-20)

$0.9(\beta_c\beta_l f_c + 2\alpha\rho_v\beta_{cor} f_{yv}) A_{ln}$

$= 0.9 \times (0.933 \times 1.323 \times 27.5 + 2 \times 0.95 \times 0.032 \times 1.181 \times 270) \times 40048$

$= 1922 \times 10^3 \text{N} = 1922\text{kN} > F_l = 1749.9\text{kN}$

The requirements are met.

9.3 Flexural Members of Prestressed Concrete

9.3.1 Basic Concept of Balanced Load Design Method

The effect of tensioning prestressed reinforcement can be replaced by a batch of equivalent loads to concrete beam. Equivalent load generally consists of two parts: (1) compression N_p of beam from prestressed tendon in anchorage zone; (2) upward distribution force w that is perpendicular to the center line of prestressed beam caused by the curvature of curved prestressed reinforcement, as is shown in Figure 9-17(b), or a upward concentrated force caused by bending up of curved prestressed reinforcement.

The above distribution or concentrated force can partly or fully offset the external load acting on the beam, which will be illustrated by an example of a single-span simply supported beam as following.

As Figure 9-17 shows, the beam is equipped with a prestressed tendon with linear of quadratic parabola, of which the equation is

$$y = 4f[x/l - (x/l)^2] \qquad (9\text{-}51)$$

Where f——the sag of a parabola.

Thus, the bending moment equation at each section along the longitudinal direction caused by prestress N_p is also a parabola, expressed as

$$M = 4N_p f(1-x)x/l^2 \qquad (9\text{-}52)$$

Where x——distance from moment calculation section to the beam end at the left.

Thus, the equivalent w caused by M according to two derivatives of x from Eq. (9-52) can be obtained as:

$$w = d^2M/dx^2 = -8N_p f/l^2 \qquad (9\text{-}53)$$

Where "$-$" means upward direction w.

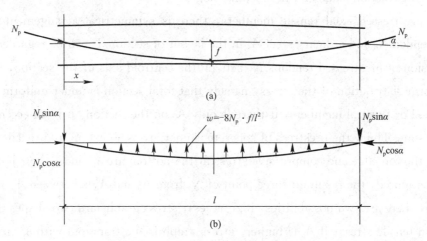

Figure 9-17 Balanced load design method

The prestress of prestressed tendons N_p on every section of beam is assumed to be equal, and $e\ (=y)$ is the eccentricity from centroid of prestressed tendon to the centroid of the beam section. Integrating Eqs. (9-51) and (9-53)

$$N_p e = w(l-x)x/2 \qquad (9\text{-}54)$$

As is shown in Eq. (9-54), if uniformly distributed loads q and w on the beam are equal, the load will all be balanced by prestress. Therefore, it is called the balanced load method put forward by Professor Lin Tongyan in 1963.

As is shown in Figure 9-17(b), the effects of prestressing force for beam include: w is the uniform load of upward effect; the horizontal force $N_{p\cos\alpha} \approx N_p$; the vertical force $N_{p\sin\alpha} \approx N_{p\tan\alpha} = 4N_p f/l$.

Therefore, under the case that uniform load q is all balanced by prestressed force w (that is, $q=w$), the vertical load of beam is zero. At this point, the beam only subjects a horizontal axis compression N_p (uniform compressive stress on section is $\sigma = N_p/A$) with no moment or vertical deflection.

If $q > w$, the bending stress on the section caused by load balance $(q-w)$ can be directly obtained by the formula of mechanics of materials.

It is to be mentioned that in order to achieve the load balance, the center lines of prestressed tendon at both ends of the beam should cross the gravity center of the section to avoid concentrated moment at the ends to break load balance of beam.

9.3.2 Stress Analysis of Flexural Members

Similar with prestressed concrete members subjected to tensile loading, the whole process from tendon jacking to a flexural member failure can also be divided into two stages: the construction stage and the loading stage.

For prestressed axial tensile members, there is symmetrical arrangement for prestressed tendon A_p and ordinary steel A_s in the layout of section. It can be regarded that the total tension of prestressed tendons N_p acts on the centroid axis of the section, leading to the uniform distribution of the stress, namely that total section is under uniform compressive state. For flexural members, if there is only A_p on the section, it is an eccentric compression state. Thus, the prestress of concrete is uneven, as shown in Figure 9-18(a) (prestress at the top edge and compressive stress at the bottom are σ'_{pc} and σ_{pc}). If both A_p and A'_p are configured, the resultant force point of N_p from A_p and A'_p is between A_p and A'_p. At this time, there are two possibilities: if A'_p is less, stress graphs are two shapes of triangle with σ'_{pc} in tensile stress; if A'_p is higher, stress graph is the trapezoid with σ'_{pc} in compressive stress less than σ_{pc}, as shown in Figure 9-18(b).

In pratical design, post-tensioned members are more commonly used. Therefore, the following chapters will mainly discuss the principles on post-tension.

Figure 9-19 shows the post-tensioned flexural members with asymmetric cross section configured prestressed tendons A_p, A'_p and ordinary steel A_s, A'_s. Compared with the stress

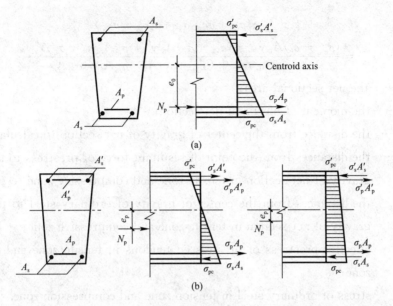

Figure 9-18 Stress of concrete on the section for prestressed flexural member
(a) Sectional stress for configuring tendons in tension zone;
(b) Sectional stress for configuring tendons in both tension and compression zone

analysis of prestressed axial tensile members corresponding to each loading stage in Chapter 9.2, the normal prestress of concrete on the cross section of prestressed flexural members σ_{pc}, effective stress of prestressed reinforcement σ_{pe}, resultant force of prestressed tendon and ordinary steel N_p, eccentric distance e_{pn} and other factors can also be obtained similarly. The equations are as the following.

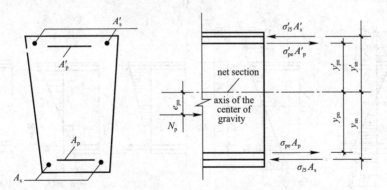

Figure 9-19 The section of posttensioned prestressed flexural member configured with both tendons and steels

1. Construction stage (Figure 9-19)

$$\sigma_{pc} = \frac{N_p}{A_n} \pm \frac{N_p e_{pn}}{I_n} y_n \tag{9-55}$$

$$N_p = \sigma_{pe} A_p + \sigma'_{pe} A'_p - \sigma_s A_s - \sigma'_s A'_s \tag{9-56}$$

$$\sigma_{pe} = \sigma_{con} - \sigma_l, \quad \sigma'_{pe} = \sigma'_{con} - \sigma'_l \tag{9-57}$$

$$\sigma_s = \alpha_E \sigma_{pc} + \sigma_{l5}, \quad \sigma'_s = \alpha_E \sigma'_{pc} + \sigma'_{l5} \tag{9-58}$$

$$e_{pn} = \frac{(\sigma_{con} - \sigma_l)A_p y_{pn} - (\sigma'_{con} - \sigma'_l)A'_p y'_{pn} - \sigma_{l5}A_s y_{sn} + \sigma'_{l5}A'_s y'_{sn}}{(\sigma_{con} - \sigma_l)A_p + (\sigma'_{con} - \sigma'_l)A'_p - \sigma_{l5}A_s - \sigma'_{l5}A_s} \tag{9-59}$$

Where A_n —— the net sectional area;

I_n —— the moment of inertia of net section;

y_n —— the distance from the center of gravity of net section to calculated fiber;

y_{pn}, y'_{pn} —— the distances from the point of resultant force of prestress to the center of gravity of net section in tensile zone and compressive zone respectively;

y_{sn}, y'_{sn} —— the distances from the center of gravity of ordinary steel to the center of gravity of net section in tensile zone and compressive zone;

$\sigma_{pe}, \sigma'_{pe}$ —— effective prestress of prestressed tendons in tensile zone and compressive zone;

σ_s, σ'_s —— stress of ordinary steel in tension zone and compression zone.

If $A'_p = 0$ on the section of member, $\sigma'_{l5} = 0$ is taken in the Eqs. (9-55)~(9-59).

Values at construction stage should be taken when using the above equations.

2. The loading stage

(1) Loading to offset the pre-compressive stress of concrete at the bottom edge

The sectional bearing moment is taken as M_0 under load (Figure 9-20c). Then the normal tensile stress of concrete at bottom edge of section is:

$$\sigma = \frac{M_0}{W_0}$$

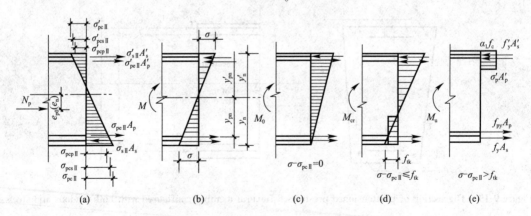

Figure 9-20 Stress changes of the section of flexural member

(a) Under prestress; (b) Under load; (c) The stress of concrete at the bottom edge of section in tension zone is zero;

(d) The concrete at the bottom edge of section in tension zone will appear cracks;

(e) The concrete at the bottom edge of section in tension zone will quit work

To offset the pre-compressive stress of concrete σ_{pcII} by above tension, that is $\sigma - \sigma_{pcII} = 0$, M_0 can be obtained as:

$$M_0 = \sigma_{pc\text{II}} W_0 \qquad (9\text{-}60)$$

Where W_0——elastic resistance moment of transformed section at the tensile edge.

Similarly, when normal stress of concrete at the point of resultant force of prestressed tendon is zero, stress of prestressed tendons in tension and compression zone are respectively σ_{p0}, σ'_{p0}

$$\sigma_{p0} = \sigma_{con} - \sigma_l + \alpha_E \frac{M_0}{W_0} \approx \sigma_{con} - \sigma_l + \alpha_E \sigma_{pc\text{II}} \qquad (9\text{-}61)$$

$$\sigma'_{p0} = \sigma'_{con} - \sigma'_l + \alpha_E \sigma_{pc\text{II}} \qquad (9\text{-}62)$$

(2) Loading to cracking in tensile zone

The cracking moment M_{cr} is the moment that causes the first crack at the extreme precompressed concrete fiber. Therefore, we have

$$M_{cr} = M_0 + \overline{M}_{cr} = \sigma_{pc\text{II}} W_0 + \gamma f_{tk} W_0 = (\sigma_{pc\text{II}} + \gamma f_{tk}) W_0$$

That is

$$\sigma = \frac{M_{cr}}{W_0} = \sigma_{pc\text{II}} + \gamma f_{tk} \qquad (9\text{-}63)$$

Where γ——coefficient considering the plasticity of concrete in the tension zone.

(3) Loading to failure

When there are vertical cracks in tensile zone, concrete on the cracking section in tensile zone will quit work and tensile reinforcement will bear all the tension. When the section goes into phase III, tensile reinforcement will yield until the failure, and stress on the cross section is similar to the bearing capacity of normal section of reinforced concrete flexural members in Chapter 3, as well as the calculation methods.

9.3.3 Design of Prestressed Flexural Members

Design of a prestressed concrete flexural member normally includes the design of normal sections, the check on crack resistance of normal sections, the design of inclined sections based on shearing capacities, the crack resistance of inclined sections and the check on the construction stage.

1. Design of normal sections

(1) Sectional stress at failure stage

The design for a prestressed normal section generally follows the same principles with the design of an ordinary non-prestressed section. But there are several special issues to be noted. In the following part, these special issues will be analyzed and the design method will be presented.

1) The relative height of compression zone at balanced failure

The tendon stress corresponding to a state at which the stress of the concrete at the very location of the tendon equals zero is defined as σ_{p0}, and the corresponding tendon strain is defined as $\varepsilon_{p0} = \sigma_{p0}/E_s$. At limiting damage, stress of prestressed tendons is up to tensile strength design values f_{py}, so stress increment of prestressed tendons in tension zone is $f_{py} - \sigma_{p0}$ and the strain increment is $(f_{py} - \sigma_{p0})/E_s$. According to plane-section assumption, relatively balanced depth in compression zone ξ_b can be determined by geometry in Figure 9-21.

$$\frac{x_c}{h_0} = \frac{\varepsilon_{cu}}{\varepsilon_{cu} + \dfrac{f_{py} - \sigma_{p0}}{E_s}} \tag{9-64}$$

At limiting damage, the balanced depth of compression zone is x_b, then $x = x_b = \beta_1 x_c$, taking into the above equation

$$\frac{x_b}{\beta_1 h_0} = \frac{\varepsilon_{cu}}{\varepsilon_{cu} + \dfrac{f_{py} - \sigma_{p0}}{E_s}} \tag{9-65}$$

That is

$$\xi_b = \frac{x_b}{h_0} = \frac{\beta_1}{1 + \dfrac{f_{py} - \sigma_{p0}}{E_s \varepsilon_{cu}}} \tag{9-66}$$

For prestressed reinforcement with no-yield point (like steel wire, etc.), according to the definition of conditional yield point, as is shown in Figure 9-22, the tensile strain of prestressed tension up to condition yield point is

$$\varepsilon_{py} = 0.002 + \frac{f_{py} - \sigma_{p0}}{E_s}$$

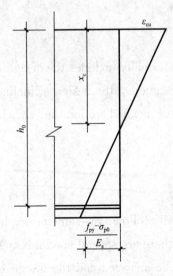

Figure 9-21 The relative height

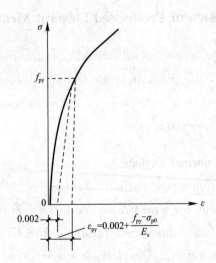

Figure 9-22 Tensile strain of the steel with conditional yielding point

Eq. (9-66) can be rewritten, then

$$\xi_b = \frac{\beta_1}{1 + \dfrac{0.002}{\varepsilon_{cu}} + \dfrac{f_{py} - \sigma_{p0}}{E_s \varepsilon_{cu}}} \quad (9\text{-}67)$$

Where σ_{p0}——stress of prestressed tendon when normal stress of concrete at the point of resultant force in longitudinal tension zone is zero.

If there are several layers of tendons or rebars, each of them will have its own ξ_b value by separate calculations, among which the smallest value ξ_b is usually used.

2) Stress in tendons or rebars at arbitrary layer

The ith tendon is taken as an example. The position of the tendon measured from the extreme compression fiber of concrete is h_{0i} and the stress of the tendon is σ_{pi}. From Figure 9-23, we have

$$\sigma_{pi} = E_s \varepsilon_{cu} \left(\frac{\beta_1 h_{0i}}{x} - 1 \right) + \sigma_{p0i} \quad (9\text{-}68)$$

If the steel bar is not prestressed, the rebar stress at ultimate state can be abtained as:

$$\sigma_{si} = E_s \varepsilon_{cu} \left(\frac{\beta_1 h_{0i}}{x} - 1 \right) \quad (9\text{-}69)$$

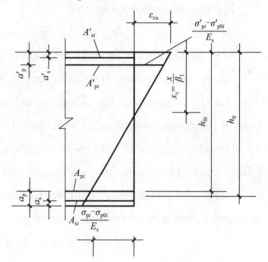

Figure 9-23　The calculation of stress σ_{pi} of tendon

If the approximate method (linear interpolation) is used, the above two equations become:

$$\sigma_{pi} = \frac{f_{py} - \sigma_{p0i}}{\xi_b - \beta_1} \left(\frac{x}{h_{0i}} - \beta_1 \right) + \sigma_{p0i} \quad (9\text{-}70)$$

$$\sigma_{si} = \frac{f_y}{\xi_b - \beta_1} \left(\frac{x}{h_{0i}} - \beta_1 \right) \quad (9\text{-}71)$$

The above two equations are commonly used in the design calculations. In these equations, σ_{pi}, σ_{si} are the stresses in the tendon and the rebar of the ith layer, respectively, with positive value for tension; h_{0i} is the distance between the center of the ith layer steel and the extreme compression fiber of concrete; x is the height of the equivalent rectangular stress block acting on the compression zone; and σ_{p0i} is the tendon stress of the ith layer corresponding to a state at which the stress in the concrete located at the center of the tendon is zero.

The above two equations are applicable only when the steel has not yielded. That is, in the case of a tendon, the following should be satisfied:

$$\sigma_{p0i} - f'_{py} \leqslant \sigma_{pi} \leqslant f_{py} \quad (9\text{-}72)$$

For rebar, the following should be satisfied:

$$f'_y \leqslant \sigma_{si} \leqslant f_y \qquad (9\text{-}73)$$

3) Calculation of stress in compression prestressed tendons (σ'_{pe})

With the increasing of load, compressive stress and compressive strain of concrete in the center of gravity of tendon A'_p increase and tensile stress in tendon A'_p reduces. Thus when the section reaches the damage state, the stress of A'_p can still be tensile or just changs into compressive stress, yet the value σ'_{pe} does not reach the design value for compressive strength f'_{py}.

For posttensioned structures

$$\sigma'_{pe} = (\sigma'_{con} - \sigma'_l) + \alpha_E \sigma'_{pc p\,II} - f'_{py} = \sigma'_{p0} - f'_{py} \qquad (9\text{-}74)$$

(2) Calculation for flexural bearing capacities of normal sections

At the ultimate limit state, the tension steels in an under-reinforced section will yield. That is, the tendons and the rebars in the tension zone will yield, as well as the normal rebars in the compression zone. The rebars A_s and A'_s can all reach their yield strength at failure. However, stress of tendons in compression zone should be calculated by Eq. (9-74) at failure stage. Therefore, for prestressed concrete flexural members with rectangular section or T-section of which flange is in tensile side (Figure 9-24), calculated formulae for flexural bearing capacities of normal sections are

$$\alpha_1 f_c bx = f_y A_s - f'_y A'_s + f_{py} A_p + (\sigma'_{p0} - f'_{py}) A'_p \qquad (9\text{-}75)$$

$$M \leqslant M_u = \alpha_1 f_c bx \left(h_0 - \frac{x}{2}\right) + f'_y A'_s (h_0 - a'_s) \\ - (\sigma'_{p0} - f'_{py}) A'_p (h_0 - a'_p) \qquad (9\text{-}76)$$

The height x of concrete in compression zone should require the following conditions:

$$x \leqslant \xi_b h_0 \qquad (9\text{-}77)$$

$$x \geqslant 2a' \qquad (9\text{-}78)$$

According to *Code for Design of Concrete Structures*, the following should be met:

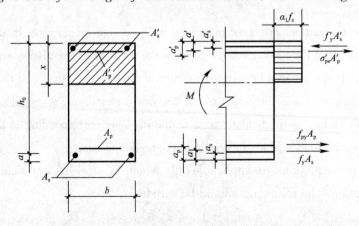

Figure 9-24 Calculation of normal sectional bearing capacity for rectangular-section flexural member

$$M_u \geqslant M_{cr} \qquad (9\text{-}79)$$

Where M ——design value of bending moment;

M_u ——design value of bending bearing capacity in normal section;

M_{cr} ——cracking moment value in normal section;

A_s, A_s' ——sectional area for longitudinal ordinary steels in tensile zone and compression zone, respectively;

A_p, A_p' ——sectional areas for longitudinal tendons in tensile zone and compressive zone, respectively;

h_0 ——the effective height of section;

b ——the width of the rectangular section or width of the web of inverted T-section;

α_1 ——when the concrete strength is not more than C50, $\alpha_1 = 1.0$ is taken; when the concrete strength is C80, $\alpha_1 = 0.94$ is taken; α_1 is considered by linear interpolation method for concrete strength between C50~C80;

a' ——the distance between the resultant of the tendons and the rebars in the compression zone and the edge of the compression zone, if there is no tendon in the compression zone or the tendon stress σ_p' is tensile, a' should be replaced by a_s';

a_s', a_p' ——the positions of the resultants of rebars and tendons in the compression zone, respectively, measured as distances from the edge of the compression zone.

In case that $x < 2a'$, the flexural strength of the section should be obtained as follows:
If σ_{pe}' is in tension, $x = 2a_s'$ is taken, as shown in Figure 9-25.

$$M \leqslant M_u = f_{py}A_p(h - a_p - a_s') + f_y A_s(h - a_s - a_s') \\ + (\sigma_{p0}' - f_{py}')A_p'(a_p' - a_s') \qquad (9\text{-}80)$$

Where a_s, a_p ——the distance from rebar, tendon in tension zone to the tensile edge of section.

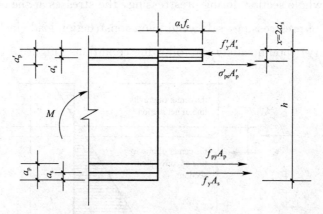

Figure 9-25 The calculated diagram of the vertical section for rectangular-section prestressed flexural member when $x < 2a'$

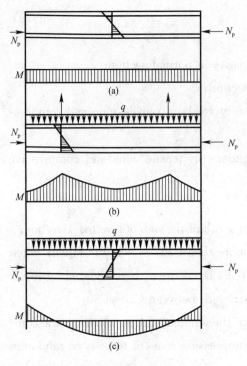

Figure 9-26 Prestressed concrete flexural member

(a) The manufacture stage; (b) The lifting stage; (c) The service stage

2. Checking at the construction stage

For prestressed flexural members, the stress at the construction stage of the production, transportation and installation is different from the loading stage. There is an eccentric compression on the section during production, when compression stress is at the bottom edge and tension stress is at the top edge (Figure 9-26a). Generally, there is a distance between the support at service and the supports during transportation and installation, and negative moments will be caused by cantilever parts in two ends, taking superposition with negative moments of eccentric pre-compression (Figure 9-26b). At the top edge of the section, if tensile stress of concrete exceeds the tensile strength of the concrete, cracks will appear, growing with time. At the bottom edge of the section, if compression of concrete is too great, longitudinal cracks will also appear. Such cracks will reduce the stiffness of the member and affect the crack resistance during service of the member. For these reasons, the construction stage should be checked under all critical conditions.

For members in which cracking is not allowed during the process of construction including manufacture, transportation and lifting, or for members that are subjected to compression over the whole section during prestressing, the stresses at the edges of the section due to the actions of prestressing, self-weight and construction load (including dynamic coefficient if necessary) should satisfy the following conditions (Figure 9-27):

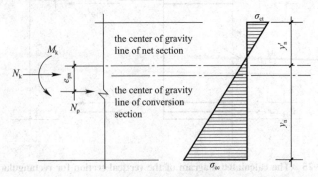

Figure 9-27 Checking in construction stage for posttensioned prestressed concrete member

$$\sigma_{ct} \leqslant f'_{tk} \quad (9-81)$$

$$\sigma_{cc} \leqslant 0.8 f'_{ck} \quad (9-82)$$

Where σ_{cc}, σ_{ct}——the compressive and tensile stresses at the edges of control sections corresponding to the construction stage;

f'_{tk}, f'_{ck}——the tensile and compressive strength of the concrete at construction corresponding to the concrete cube strength f'_{cu}, respectively.

The terms σ_{cc} and σ_{ct} are calculated by using the following equations:

$$\left.\begin{array}{c}\sigma_{cc}\\ \sigma_{ct}\end{array}\right\} = \sigma_{pc} + \frac{N_k}{A_0} \pm \frac{M_k}{W_0} \quad (9-83)$$

Where σ_{pc}——the normal stress in concrete due to prestressing (positive for compression);

N_k, M_k——the axial force and moment on the control section due to the self-weight and construction load;

W_0——the section modulus for the corresponding section edges.

3. Deformation checking for flexural members

Deflection of prestressed flexural member is formed by the superposition of two parts: one part f_{1l} is generated by loading; the other part f_{2l} is generated by arch of prestressing.

(1) Deflection of components under load f_{1l}

Deflection f_{1l} can be calculated according to the method of mechanics of materials, that is,

$$f_{1l} = S\frac{Ml^2}{B} \quad (9-84)$$

Bending stiffness B should be calculated according to the following conditions, respectively:

1) Short-term stiffness under standard combination of loads can be calculated by the following formula:

For members that cracks are not allowed at loading stage, the stiffness is:

$$B_s = 0.85 E_c I_0 \quad (9-85)$$

Where E_c——the elastic modulus of concrete;

I_0——inertia moment of conversion section;

0.85——the stiffness reduction factor, taking into account the plastic deformation before cracks occur in tension zone.

For members that cracks are allowed at load stage, the stiffness is

$$B_s = \frac{0.85 E_c I_0}{k_{cr} + (1 - k_{cr})w} \quad (9-86)$$

$$k_{cr} = \frac{M_{cr}}{M_k} \quad (9-87)$$

$$\omega = \left(1 + \frac{0.21}{\alpha_E \rho}\right)(1 + 0.45\gamma_f) - 0.7 \qquad (9\text{-}88)$$

$$M_{cr} = (\sigma_{pcII} + \gamma f_{tk})W_0 \qquad (9\text{-}89)$$

Where k_{cr} ——ratio of cracking moment M_{cr} of normal section of prestressed concrete flexural member to moment M_k under standard combination of loads, when $k_{cr} > 1.0$, $k_{cr} = 1.0$ is taken;

γ ——plastic effect coefficient of section modulus for concrete members;

σ_{pcII} ——pre-compressive stress of concrete caused by prestress at the edge of crack resistance after deducting all prestress losses;

α_E ——ratio of elastic modulus of steel and elastic modulus of concrete, $\alpha_E = E_s/E_c$;

ρ ——reinforcement ratio of longitudinal tensile steel, $\rho = (\alpha_1 A_p + A_s)/(bh_0)$; for grouting post-tensioned prestressed tendon, $\alpha_1 = 1.0$ is taken;

γ_f ——ratio of sectional area of tension flange with effective sectional area of the web; $\gamma_f = (b_f - b)h_f/bh_0$, where b_f and h_f are the width and height of tensile flange, respectively.

For members with cracks in tension zone in pretensioning, B_s should reduce 10%.

2) The stiffness that considers the long-term effect of prestressing according to standard combinations of loads can be calculated by the sectional stiffness B equation in Chapter 8, where $\theta = 2.0$ is taken and B_s is calculated by Eq. (9-85) or (9-86).

(2) Inverted arch caused by prestress f_{2l}

Inverted arch is caused by eccentricity of prestressed concrete members e_p under the action of total pre-compression N_p, the value of which can be calculated according to the formula of structural mechanics, namely by simply supported beam with bending moments $(N_p e_p)$ at both ends. The beam span is defined as l and the sectional bending stiffness is B, then

$$f_{2l} = \frac{N_p e_p l^2}{8B} \qquad (9\text{-}90)$$

Where N_p, e_p and B take different values under the following different situations:

1) Inverted arch caused by prestress for members

According to standard combinations of loads, $B = 0.85 E_c I_0$, when N_p and e_p are calculated by the case of deducting the first prestressing loss value. Posttensioned members are defined as N_{pI}, e_{pnI}.

2) Prestressing inverted arch value at loading stage

With the long-term effect of prestress at loading stage, creep deformation of concrete in pre-compression zone increases inverted arch values of beam. Therefore, the stiffness can be calculated by $B = E_c I_0$ taking into account the long-term effect of pre-compressive stress. At this time, N_p and e_p should be calculated after deducting all the prestress losses, i. e. N_{pII} and e_{pnII} for posttensioned members. For simplified calculation, inverted arch value can multiply the magnification factor 2.0.

(3) The calculation of deflection

Inverted arch caused by deflection under standard combinations of loads deducting the deflection from prestress, is the final deflection of prestressed flexural members:

$$f = f_{1l} - f_{2l} \leqslant [f] \qquad (9\text{-}91)$$

Where $[f]$——limited value of deflection.

In total, steps for design of prestresed flexural members are:

1) The section dimension, as well as the strengths and elastic modulus of the concrete, the tendons and the rebars are to be determined. The strength of concrete at the time of prestress transfer, the control stress of the tendons and the construction method (pretensioning or posttensioning) should be selected. The initial section area of the tendons and the rebars are chosen, followed by looking up the structural importance coefficient. The internal forces of the structure is calculated.

2) The prestress loss σ_l is calculated.

3) The effective prestress σ_{peII} is calculated.

4) The bearing capacities of normal sections at ultimate limit state are calculated. If the bearing capacity requirement is not satisfied, the procedure should go back to step 1 to adjust the parameters and recalculate.

5) The crack resistance and crack width (if applicable) of the normal sections under serviceability conditions are checked. If any of the requirements are not satisfied, the procedure should go back to step 1.

6) The shear capacity at ultimate limit state is checked. If this is not satisfied, the parameters should be adjusted for recalculation.

7) The resistance to inclined cracking under serviceability conditions is to be checked. If this is not satisfied, the parameters should be adjusted for recalculation.

8) The requirements of deflections should be met. Or else, the parameters are to be adjusted for recalculation;

9) The construction stage should be checked including manufacture, transportation, lifting and erection. If any of these are not satisfied, parameters should adjust for recalculation.

9.4 Detail Requirements of Prestressed Concrete Members

1. Sectional types and dimensions

Prestressed axial tension members are usually in square or rectangular sections. Prestressed flexural members can use T-shaped, I-shaped and box-section.

In order to expediently arrange tendons and be sure that there is enough compressive ability in pre-compression zone, I-shaped section with asymmetric top and bottom flanges can be designed. The width of bottom tension flange can be narrower than the top flange, but the thickness is larger than the top flange.

The dimensions of sections can be varied along the longitudinal axis. For example, it is I-shaped in the middle span, while it is generally designed into rectangular section at the ends so that it can bear large shear force near the end and leave sufficient positions for anchorage.

Because of the great crack resistance and stiffness for prestressed members, the sectional size can be smaller than ordinary reinforced member. For prestressed concrete flexural member, its section height can be $h = l/20 \sim l/14$, minimal $l/35$ (l is the span), generally preferable to about 70% of the height of reinforced concrete beams. Flange width is generally preferable to $h/3 \sim h/2$ with thickness of $h/10 \sim h/6$, and web width is as small as possible with a desirable value of $h/15 \sim h/8$.

2. Arrangement for prestressed longitudinal reinforcement and additional vertical steel at the end

Straight line style: with small span or lower loads, the most common style is the line style, as is shown in Figure 9-28(a).

Curved or polygonal line style: with larger spans or loads, the prestressed reinforcemcent can be arranged into a curved shape (Figure 9-28b) or polygonal line style (Figure 9-28c) by posttensioning in construction, such as prestressed concrete roof beams, crane girders and other components. In order to withstand the main tensile stresses near the end and protect against cracks caused by prestressing in pretension zone or horizontal cracks along the central section at the end of member, some tendons are bent up near the support with uniform arrangement.

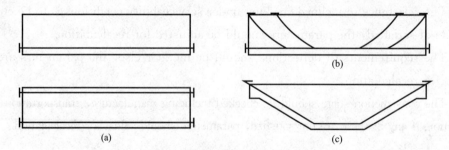

Figure 9-28 The distribution of tendons
(a) Straight line shape; (b) Curved line shape; (c) Polygonal line shape

When the tendons at the end of component are needed to be arranged at the upper or lower of section, additional vertical steel mesh should be set up. Closed stirrups or other forms of structural steels and rebars should be adopted within $0.2h$ at the end to prevent the end cracks, sectional area of which should meet the following requirements:

$$A_{sv} \geqslant \frac{T_s}{f_{yv}} \qquad (9\text{-}92)$$

$$T_s = \left(0.25 - \frac{e}{h}\right)P \qquad (9\text{-}93)$$

When $e > 0.2h$, structural steel can be arranged according to the real conditions.

Where T_s——the tension force on the anchor end;

P——resultant force design value acting above or below the center of sectional gravity line at the end of member;

e——the distance from the point of resultant force of tendons above or below the center of sectional gravity line to the edge of section;

f_{yv}——tensile strength design values for additional reinforcement, taken according to the Table 1-11 in Appendix 1.

When there are both ten dons in the upper and lower of the end section, the total sectional area of additional vertical steels should take the larger value by comparing the results from the resultant forces of prestress in upper section and lower section, respectively.

The end section should also calculate the cracking resistance steel along the transverse direction, forming steel mesh pieces with the above vertical steels.

3. Arrangement of longitudinal steel

For prestressed components, besides prestressed tendon, sufficient steels in pretension zone should be arranged in order to prevent cracks in pretension zone in the process of shrinkage of concrete, temperature difference, prestressing in construction, and reduce crack width in the process of production, stacking, transporting and lifting.

In pretension and pre-compression zone for post-tensioned prestressed concrete members, vertically ordinary structural steel should be installed; in the zone of prestressed tendon bending, stirrups should be densely arranged or normal reinforcement mesh should be arranged along the inside of bending in order to strengthen concrete in the sections of bent rebars.

For flexural members with tendons all bent up at the end of components or pretensioned members of straight line reinforcement, when welded between the ends of members and the lower supporting structure, the negative effects should be considered because of shrinkage, creep and temperature variations of concrete and enough regular longitudinal structural steels should be configured in the positions of members where cracks may appear.

4. Net space of steel wires and steel strands

Net space of pretensioned prestressed tendon should be based on concrete casting, prestressing and tendon anchorage requirements. Net space between prestressed tendons should not be smaller than 2.5 times of the nominal diameter and 1.25 times of the maximum grain size of coarse aggregate of concrete, which should also comply with the following requirements:

The diameter should be not less than 15mm for prestressed steel wire;

The diameter should be not less than 20mm for three steel strands and not less than 25mm for seven steel strands.

5. Duct spacing for posttensioned prestressed tendons

(1) The clear horizontal spacing between the ducts is suggested to be not less than 50mm and 1.25 times the diameter of coarse aggregates. The concrete cover to the duct (the clear distance between the duct and the edge of the member) is preferred to be at least 30mm and should not be less than half of the duct diameter.

(2) In a frame beam, the moments are usually negative at supports and positive at mid-spans. As a result, the tendons are usually curved accordingly. The ducts for such tendons should have a clear spacing, that is, at least 1.5 times the duct diameter in the horizontal direction and one time the duct diameter in the vertical direction with a preferred diameter 1.25 times larger than the diameter of coarse aggregates. The concrete cover to the ducts is suggested to have a clearance of at least 40mm at the sides of the beam and at least 50mm at the bottom of the beam.

(3) The diameter of a duct should be greater at least 10~15mm than the outer diameter of the prestressing wire bundle or prestressing strand, or any part of the anchorage in a duct. And the cross-sectional area of the duct is preferred to be 3~4 times that of the tendon.

(4) Grouting holes or vent should be set in the ends or at mid-span of components and the spacing of these holes should not be greater than 12m.

(5) When camber is adopted, the ducts should also be cambered to avoid secondary stresses which are not considered in the design.

6. Anchor

The anchorages should be reliable and conform to the specifications of the current national standards.

7. Measures for reinforcing at the end of member

For single configured tendons of pretensioned prestressed concrete members, spiral reinforcement should be set at the ends; for distributed multiple tendons, 3~5 pieces of steel mesh vertical with prestressed tendon should be set within $10d$ of the member ends and not less than 100mm length (d is the nominal diameter of prestressing tendon).

The end size of posttensioned members should consider the anchor layout, size of tensioning device and requirements of local compression. The size should be increased if necessary.

Under the prestressed tendon anchorage and at the support of tensioning devices, the embedded steel plate and structural transverse reinforcement mesh or spiral reinforcement and other measures for partial hardening should be set.

For exposed metal anchor, reliable corrosion protection and fire prevention measures should be adopted.

For curved prestressed steel wire beams and steel strand beams of posttensioned prestressed concrete members, the radius of curvature is not less than 4m.

For members with polygonal line steel, the radius of curvature in the bending of tendon can reduce properly.

Additional indirect reinforcement should be provided in this zone, which is outside the zone determined by local bearing requirements. This zone has a height of $2e$ and a length of $3e$ (the length is not greater than $1.2h$), where e is the distance from the resultant tendon forces (due to local compression) to the edge of the member and h is the section height at the end (Figure 9-29). The corresponding volumetric reinforcement ratio should also be no less than 0.5%. The area of reinforcement can be calculated by

$$A_{sb} \geqslant 0.18\left(1-\frac{l_l}{l_b}\right)\frac{P}{f_{yv}} \qquad (9-94)$$

Where P and f_{yv} are the same meanings with Eqs. (9-92) and (9-93), l_l, l_b are side length or diameter of A_l and A_b along the height, respectively. A_l and A_b can be calculated based on Chapter 9.1.10.

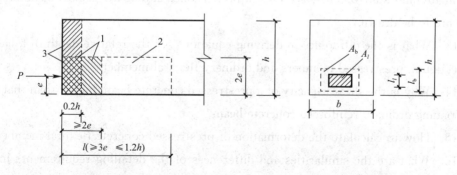

Figure 9-29 The range of additional reinforcement to prevent cracks at the ends

1—the zone of indirect reinforcement; 2—the additional reinforcement zone for preventing split;
3—the additional reinforcement zone for preventing cracks at the ends

Questions

9-1　What are prestressed concrete structures?

9-2　Why high-strength concrete and high-strength reinforcement should be used in prestressed concrete structures?

9-3　What is a fully prestressed concrete structure? What is a partially prestressed concrete structure? What is the degree of prestressing?

9-4　How many kinds of prestress losses are there in prestressed concrete structures? How do we calculate them? And how do we reduce them?

9-5　What are the stress transfer mechanisms of different types of anchorages?

9-6　What are pretensioned and posttensioned concrete structures?

9-7　What are the factors that will influence the allowable value of control stress? Why are these factors important?

9-8　What are the requirements on materials in prestressed concrete structures?

9-9　What is the transfer length of prestress? How do we check the bearing capacity at the ends of posttensioned members?

9-10　The internal force diagrams, stress diagrams at releasing state, cracking state and ultimate limit state on cross sections of pretensioned and posttensioned members under axial tension are to be drawn.

9-11　Where are tendons usually placed in flexural members? The tendons are to be illustrated in a simply supported beam, a simply supported beam with cantilever part and a two-span continuous beam, respectively.

9-12　The internal force diagrams, stress diagrams at releasing state, cracking state and ultimate limit state on cross sections of pretensioned and posttensioned members under flexure are to be drawn.

9-13　What is the difference in deriving equations for the relative depth of a balanced cross section in prestressed members and ordinary flexural members?

9-14　Why is the shear capacity of a prestressed concrete beam higher than that of the corresponding ordinary reinforced concrete beam?

9-15　How to calculate the deformation of prestressed concrete flexural members?

9-16　What are the similarities and differences of the detailing requirements in prestressed concrete members and ordinary reinforced concrete members?

Exercises

9-1 A pretensioned concrete member is under axial tension. The cross-sectional dimensions of the member are 200mm×200mm. C40 concrete is used, and 9 $\phi^H 9$ tendons are provided. The control stress is $\sigma_{con}=1000\text{N/mm}^2$ and $f_{py}=1110\text{N/mm}^2$. There is no ordinary reinforcement. The first and second batch of prestress losses are $\sigma_{l\text{I}}=68\text{N/mm}^2$ and $\sigma_{l\text{II}}=52\text{N/mm}^2$, respectively. The calculations are required: (1) the prestress σ_c in construction; (2) the load needed to offset the normal compressive stress in concrete; (3) the cracking load; (4) the ultimate capacity of the member.

9-2 A pretensioned concrete beam is subjected to a bending moment of $M=10\text{kN}\cdot\text{m}$. The sectional dimensions are $b\times h=150\text{mm}\times 300\text{mm}$. Tendons of $3\phi^{HM}9$ are provided in the tension zone. The control stress is $\sigma_{con}=0.7f_{ptk}$, $f_{ptk}=1270\text{N/mm}^2$. Prestress losses are $\sigma_{l1}=10.5\text{N/mm}^2$, $\sigma_{l3}=0$, $\sigma_{l4}=19\text{N/mm}^2$, and $\sigma_{l5}=53\text{N/mm}^2$. The concrete grade is C40. The flexural capacity of the beam is to be checked; also, the beam strengths in the construction stage are to be checked when tendons are released, at which the concrete has attained 85% of the design strength.

Appendix 1
Indexes of Mechanical Properties of Materials as Specified in *Code for Design of Concrete Structures* GB 50010—2010

Standard value of axial compressive strength of concrete (N/mm²)   Table 1-1

Strength	Strength grades of concrete													
	C15	C20	C25	C30	C35	C40	C45	C50	C55	C60	C65	C70	C75	C80
f_{ck}	10.0	13.4	16.7	20.1	23.4	26.8	29.6	32.4	35.5	38.5	41.5	44.5	47.4	50.2

Standard value of axial tensile strength of concrete (N/mm²)   Table 1-2

Strength	Strength grades of concrete													
	C15	C20	C25	C30	C35	C40	C45	C50	C55	C60	C65	C70	C75	C80
f_{tk}	1.27	1.54	1.78	2.01	2.20	2.39	2.51	2.64	2.74	2.85	2.93	2.99	3.05	3.11

Design value of axial compressive strength of concrete (N/mm²)   Table 1-3

Strength	Strength grades of concrete													
	C15	C20	C25	C30	C35	C40	C45	C50	C55	C60	C65	C70	C75	C80
f_c	7.2	9.6	11.9	14.3	16.7	19.1	21.1	23.1	25.3	27.5	29.7	31.8	33.8	35.9

Design value of axial tensile strength of concrete (N/mm²)   Table 1-4

Strength	Strength grades of concrete													
	C15	C20	C25	C30	C35	C40	C45	C50	C55	C60	C65	C70	C75	C80
f_t	0.91	1.10	1.27	1.43	1.57	1.71	1.80	1.89	1.96	2.04	2.09	2.14	2.18	2.22

Modulus of elasticity of concrete ($\times 10^4$ N/mm²)   Table 1-5

Strength grades of concrete	C15	C20	C25	C30	C35	C40	C45	C50	C55	C60	C65	C70	C75	C80
E_c	2.20	2.55	2.80	3.00	3.15	3.25	3.35	3.45	3.55	3.60	3.65	3.70	3.75	3.80

Note: 1. When there is a reliable test basis, the elastic modulus can be determined according to the measured data;
2. When the concrete is mixed with a large amount of mineral admixture, the elastic modulus can be determined according to the measured data under measured age.

Modification coefficient of compressive fatigue strength of concrete γ_ρ Table 1-6

ρ_c^f	$0 \leqslant \rho_c^f < 0.1$	$0.1 \leqslant \rho_c^f < 0.2$	$0.2 \leqslant \rho_c^f < 0.3$	$0.3 \leqslant \rho_c^f < 0.4$	$0.4 \leqslant \rho_c^f < 0.5$	$\rho_c^f \geqslant 0.5$
γ_ρ	0.68	0.74	0.80	0.86	0.93	1.00

Modification coefficient of tensile fatigue strength of concrete γ_ρ Table 1-7

ρ_c^f	$0 \leqslant \rho_c^f < 0.1$	$0.1 \leqslant \rho_c^f < 0.2$	$0.2 \leqslant \rho_c^f < 0.3$	$0.3 \leqslant \rho_c^f < 0.4$	$0.4 \leqslant \rho_c^f < 0.5$
γ_ρ	0.63	0.66	0.69	0.72	0.74
ρ_c^f	$0.5 \leqslant \rho_c^f < 0.6$	$0.6 \leqslant \rho_c^f < 0.7$	$0.7 \leqslant \rho_c^f < 0.8$	$\rho_c^f \geqslant 0.8$	—
γ_ρ	0.76	0.80	0.90	1.00	—

Note: For the concrete components directly subjected to fatigue load, curing temperature should not be higher than 60℃ under steam curing.

Fatigue deformation modulus of concrete ($\times 10^4 \text{N/mm}^2$)   Table 1-8

Strength grades	C30	C35	C40	C45	C50	C55	C60	C65	C70	C75	C80
E_c^f	1.30	1.40	1.50	1.55	1.60	1.65	1.70	1.75	1.80	1.85	1.90

Standard value of strength of steel bar (N/mm²)   Table 1-9

Type of steel bar	Symbol	Nominal diameter d (mm)	Standard value of yield strength f_{yk}	Standard value of limit strength f_{stk}
HPB300	Φ	6~22	300	420
HRB335 HRBF335	Φ Φ^F	6~50	335	455
HRB400 HRBF400 RRB400	Φ Φ^F Φ^R	6~50	400	540
HRB500 HRBF500	Φ Φ^F	6~50	500	630

Standard value of strength of prestressed tendon (N/mm²)   Table 1-10

Type of prestressed tendons		Symbol	Nominal diameter d(mm)	Standard value of yield strength f_{pyk}	Standard value of limit strength f_{ptk}
Median-strength prestressed wire	Plane surface wire Spiral rib wire	Φ^{PM} Φ^{HM}	5, 7, 9	620	800
				780	970
				980	1270
Prestressed twisted steel bar	Twisted surface	Φ^T	18, 25, 32, 40, 50	785	980
				930	1080
				1080	1230
Stress-relieved wire	Plane surface wire Spiral rib wire	Φ^P Φ^H	5	—	1570
				—	1860
			7	—	1570
			9	—	1470
				—	1570

continued

Type of prestressed tendons		Symbol	Nominal diameter d(mm)	Standard value of yield strength f_{pyk}	Standard value of limit strength f_{ptk}
Strand	1×3	Φ^S	8.6,10.8,12.9	—	1570
				—	1860
				—	1960
	1×7		9.5,12.7, 15.2,17.8	—	1720
				—	1860
				—	1960
			21.6	—	1860

Note: When the standard value of ultimate strength of the steel strand is 1960N/mm² for post-tensioned prestressed reinforcement, a reliable engineering experience should be ensured.

Design value of strength of steel bar (N/mm²)　　　　Table 1-11

Type of steel bar	Design value of tensile strength f_y	Design value of compressive strength f'_y
HPB300	270	270
HRB335, HRBF335	300	300
HRB400, HRBF400, RRB400	360	360
HRB500, HRBF500	435	410

Design value of strength of prestressed tendon (N/mm²)　　　　Table 1-12

Type of prestressed tendons	Standard value of limit strength f_{ptk}	Design value of tensile strength f_{py}	Design value of compressive strength f'_{py}
Median-strength prestressed wire	800	510	410
	970	650	
	1270	810	
Stress-relieved wire	1470	1040	410
	1570	1110	
	1860	120	
Strand	1570	1110	390
	1720	1220	
	1860	1320	
	1960	1390	
Prestressed twisted steel bar	980	650	410
	1080	770	
	1230	900	

Note: When the standard value of strength of prestressed tendon does not conform to the provisions of this table, the design value of strength should be scaled adopted accordingly.

The total elongation limit of steel bar and prestressed tendon under maximum force　　Table 1-13

Type of steel bar and prestressed tendon	Steel bar			Prestressed tendon
	HPB300	HRB335, HRBF335, HRB400, HRBF400, HRB500, HRBF500	RRB400	
δ_{gt}(%)	10.0	7.5	5.0	3.5

Modulus of elasticity of steel bar and prestressed tendon ($\times 10^5 \text{N/mm}^2$)　　Table 1-14

Type of steel bar and prestressed tendon	Modulus of elasticity E_s
HPB300	2.10
HRB335, HRBF335, HRB400, HRBF400, HRB500, HRBF500, RRB400, prestressed twisted steel bar	2.00
Stress-relieved wire, median-strength prestressed wire	2.05
Strand	1.95

Note: The measured modulus of elasticity can be adopted if necessary.

Fatigue stress limit of steel bar (N/mm^2)　　Table 1-15

Fatigue stress ratio ρ_s^f	Fatigue stress limit Δf_y^f	
	HRB335	HRB400
0	175	175
0.1	162	162
0.2	154	156
0.3	144	149
0.4	131	137
0.5	115	123
0.6	97	106
0.7	77	85
0.8	54	60
0.9	28	31

Note: When the longitudinal tensile steel bar adopts flash contact welding connection, the fatigue stress limit value at joint should be accessed according to the value of the table multiplied by 0.8.

Fatigue stress limit of prestressed tendon (N/mm^2)　　Table 1-16

Fatigue stress ratio ρ_p^f	Strand $f_{ptk}=1570$	Stress-relieved wire $f_{ptk}=1570$
0.7	144	240
0.8	118	168
0.9	70	88

Note: 1. When ρ_p^f is not less than 0.9, there is no need for the fatigue test of prestressed tendon;
2. When there is sufficient basis, the fatigue stress limit at the table can make appropriate adjustments.

Appendix 2
The Nominal Diameter, Nominal Cross-Sectional Area and Theoretical Weight of Nominal Steels

The nominal diameter, nominal cross-sectional area and theoretical weight of nominal steel   Table 2-1

Nominal diameter (mm)	Area of steel with number of (mm^2)									Mass of single steel bar (kg/m)
	1	2	3	4	5	6	7	8	9	
6	28.3	57	85	113	142	170	198	226	255	0.222
8	50.3	101	151	201	252	302	352	402	453	0.395
10	78.5	157	236	314	393	471	550	628	707	0.617
12	113.1	226	339	452	565	678	791	904	1017	0.888
14	153.9	308	461	615	769	923	1077	1231	1385	1.21
16	201.1	402	603	804	1005	1206	1407	1608	1809	1.58
18	254.5	509	763	1017	1272	1527	1781	2036	2290	2.00(2.11)
20	314.2	628	942	1256	1570	1884	2199	2513	2827	2.47
22	380.1	760	1140	1520	1900	2281	2661	3041	3421	2.98
25	490.9	982	1473	1964	2454	2945	3436	3927	4418	3.85(4.10)
28	615.8	1232	1847	2463	3079	3695	4310	4926	5542	4.83
32	804.2	1609	2413	3217	4021	4826	5630	6434	7238	6.31(6.65)
36	1017.9	2036	3054	4072	5089	6107	7125	8143	9161	7.99
40	1256.6	2513	3770	5027	6283	7540	8796	10053	11310	9.87(10.34)
50	1963.5	3928	5892	7856	9820	11784	13748	15712	17676	15.42(16.28)

Note: The brackets are the values of the prestressed thread.

Area of bar section (mm^2) per width of slab   Table 2-2

Spacing of steel bar (mm)	Diameter (mm)											
	3	4	5	6	6/8	8	8/10	10	10/12	12	12/14	14
70	101.0	180.0	280.0	404.0	561.0	719.0	920.0	1121.0	1369.0	1616.0	1907.0	2199.0
75	94.2	168.0	262.0	377.0	524.0	671.0	859.0	1047.0	1277.0	1508.0	1780.0	2052.0
80	88.4	157.0	245.0	354.0	491.0	629.0	805.0	981.0	1198.0	1414.0	1669.0	1924.0
85	83.2	148.0	231.0	333.0	462.0	592.0	758.0	924.0	1127.0	1331.0	1571.0	1811.0
90	78.5	140.0	218.0	314.0	437.0	559.0	716.0	872.0	1064.0	1257.0	1483.0	1710.0
95	74.5	132.0	207.0	298.0	414.0	529.0	678.0	826.0	1008.0	1190.0	1405.0	1620.0

continued

Spacing of steel bar (mm)	Diameter (mm)											
	3	4	5	6	6/8	8	8/10	10	10/12	12	12/14	14
100	70.6	126.0	196.0	283.0	393.0	503.0	644.0	785.0	958.0	1131.0	1335.0	1539.0
110	64.2	114.0	178.0	257.0	357.0	457.0	585.0	714.0	871.0	1028.0	1214.0	1399.0
120	58.9	105.0	163.0	236.0	327.0	419.0	537.0	654.0	798.0	942.0	1113.0	1283.0
125	56.5	101.0	157.0	226.0	314.0	402.0	515.0	628.0	766.0	905.0	1068.0	1231.0
130	54.4	96.6	151.0	218.0	302.0	387.0	495.0	604.0	737.0	870.0	1027.0	1184.0
140	50.5	89.8	140.0	202.0	281.0	359.0	460.0	561.0	684.0	808.0	954.0	1099.0
150	47.1	83.8	131.0	189.0	262.0	335.0	429.0	523.0	639.0	754.0	890.0	1026.0
160	44.1	78.5	123.0	177.0	246.0	314.0	403.0	491.0	599.0	707.0	834.0	962.0
170	41.5	73.9	115.0	166.0	231.0	296.0	379.0	462.0	564.0	665.0	785.0	905.0
180	39.2	69.8	109.0	157.0	218.0	279.0	358.0	436.0	532.0	628.0	742.0	855.0
190	37.2	66.1	103.0	149.0	207.0	265.0	339.0	413.0	504.0	595.0	703.0	810.0
200	35.3	62.8	98.2	141.0	196.0	251.0	322.0	393.0	479.0	505.0	668.0	770.0
220	32.1	57.1	89.2	129.0	179.0	229.0	293.0	357.0	436.0	514.0	607.0	700.0
240	29.4	52.4	81.8	118.0	164.0	210.0	268.0	327.0	399.0	471.0	556.0	641.0
250	28.3	50.3	78.5	113.0	157.0	201.0	258.0	314.0	383.0	452.0	534.0	616.0
260	27.2	48.3	75.5	109.0	151.0	193.0	248.0	302.0	369.0	435.0	513.0	592.0
280	25.2	44.9	70.1	101.0	140.0	180.0	230.0	280.0	342.0	404.0	477.0	550.0
300	23.6	41.9	65.5	94.2	131.0	168.0	215.0	262.0	319.0	377.0	445.0	513.0
320	22.1	39.3	61.4	88.4	123.0	157.0	201.0	245.0	299.0	353.0	417.0	481.0

The nominal diameter, nominal cross-sectional area and theoretical weight of strand   Table 2-3

Type of strand	Nominal diameter (mm)	Nominal area of section (mm²)	Theoretical weight (kg/m)
1×3	8.6	37.7	0.296
	10.8	58.9	0.462
	12.9	84.8	0.666
1×7	9.5	54.8	0.430
	12.7	98.7	0.775
	15.2	140	1.101
	17.8	191	1.500
	21.6	285	2.237

The nominal diameter, nominal cross-sectional area and theoretical weight of wire   Table 2-4

Nominal diameter (mm)	Nominal area of section (mm²)	Theoretical weight (kg/m)
5.0	19.63	0.154
7.0	38.48	0.302
9.0	63.62	0.499

Appendix 3
Environmental Categories of Concrete Structures

Environmental categories in which the concrete structures are located Table 3-1

Environmental category		Condition
1		Indoor dry environment; Permanent non-erosive hydrostatic submerged environment
2	a	Indoor humid environment; Non-freezing and non-cold areas of the open air environment; Non-freezing and non-cold areas with non-erosive water and soil in direct contact; Freezing and cold areas below the freezing line with non-erosive water and soil in direct contact
2	b	Dry and wet alternating environment; Frequent changes in water level environment; Freezing and cold areas of the open air environment; Freezing and cold areas above the frozen line with non-invasive water and soil in direct contact
3	a	Freezing and cold areas in winter water level changes; By the impact of ice salt environment; Sea breeze environment
3	b	Saline soil environment; By the impact of ice salt environment; Coastal environment
4		Seawater environment
5		Affected by human or natural erosive substances

Note: (1) Indoor humid environment is the surface of the component which is often in the state of condensation or moist environment.
(2) The division of freezing and cold areas should be in conformity with the relevant provisions of *Thermal Design Code for Civil Building* GB 50176.
(3) Coastal environment and sea breeze environment should be based on local conditions, taking into account the dominant wind direction and the structure of the windward, leeward and other factors, which are determined by researches and engineering experience.
(4) The environment affected by deicing salt is based on the salt mist of the deicing salt; the deicing salt environment refers to the environment which is removed by the deicing salt solution, and the car washing room, parking lot with deicing salt, etc.
(5) Exposure to the environment refers to the environment of the concrete surface.

References

[1] Ministry of Housing and Urban-Rural Construction of the People's Republic of China. Code for Design of Concrete Structures GB 50010—2010[S]. Beijing: China Architecture & Building Press, 2010. (in Chinese)

[2] Southeast University, Tianjin University & Tongji University. The design theory for concrete structures[M]. The fifth edition. Beijing: China Architecture & Building Press, 2012. (in Chinese)

[3] Jun Zhao, Xinling Wang, Liusheng Chu, Hui Qian &Le Li. Reinforced concrete fundamentals[M]. Beijing: China Architecture & Building Press, 2015.

[4] Xianglin Gu, Xianyu Jin, Yong Zhou. Basic principles of concrete structures[M]. Shanghai: Tongji University Press, 2015.

[5] Xingwen Liang, Qingxuan Shi. The design theory for concrete structures[M]. The third edition. Beijing:China Architecture & Building Press, 2016. (in Chinese)

[6] Lieping Ye. Concrete structures[M]. The Second edition. Beijing: Tsinghua University Press, 2005. (in Chinese)

[7] Jianjing Jiang, Xinzheng Lu, Bo Jiang. Design of reinforced concrete basic members[M]. Beijing: Tsinghua University Press, 2006. (in Chinese)

[8] Zongjian Lan. The design theory for concrete structures[M]. Nanjing: Southeast University Press, 2008. (in Chinese)

[9] Pusheng Shen. The design theory for concrete structures[M]. The third edition. Beijing: Higher Education Press, 2007. (in Chinese)

[10] Jianshu Ye. Principle of structural design[M]. The third edition. Beijing: China Communications Press, 2014. (in Chinese)

[11] Yu Zhang. The design theory for concrete structures[M]. Beijing: China Architecture & Building Press, 2000. (in Chinese)

[12] Tongyan Lin, N. H. Burns. Prestressed concrete structure design[M]. The third edition. Beijing:China Railway Press, 1984.

[13] Arthur H. Nilson, David Darwin, Charles W. Dolan. Design of concrete structures[M]. Harbin Institute of Technology Press, 2015.